Water Fit to Drink

Water Fit to Drink

A guide to the hidden hazards of drinking water and what you can do to ensure a safe, good-tasting supply for the home

by Carol Keough

with special contributions and
technical assistance from
Patricia M. Nesbitt

foreword by Robert H. Harris, Ph.D.
Member, President's Council
on Environmental Quality

Illustrations by Edward Hanke

Rodale Press   Emmaus, Pa.

Printed in the United States of America on recycled paper, containing a high percentage of de-inked fiber.

Book Design by Kim E. Morrow

Library of Congress Cataloging in Publication Data

Keough, Carol.
  Water fit to drink.

  Includes bibliographical references and index.
  1. Drinking water. I. Nesbitt, Patricia M.
II. Title.
TD380.K43   363.6′1   80–13913
ISBN 0–87857–296–1 hardcover
ISBN 0–87857–297–X paperback

2 4 6 8 10 9 7 5 3 1 hardcover

2 4 6 8 10 9 7 5 3 1 paperback

36624

Contents

Foreword

John Snow's classic studies of cholera in London in the early 1850s paved the way for the development of the water purification practices that are in common use today. Snow conclusively demonstrated that the polluted River Thames, which served as a drinking water supply for much of London, was responsible for the cholera epidemics that plagued that city. Moving water supply intakes to less polluted reaches of rivers and the development of sand filtration were instrumental in reducing the toll of waterborne diseases in Europe in the years that followed Snow's work.

It was not until around the turn of the century, however, that similar steps were taken by American cities to reduce the incidence of waterborne diseases, which were among the leading causes of death at that time. Sand filters, coagulation and sedimentation basins, chlorination—these processes became standard fare at drinking water purification plants in most major United States cities in the early 1900s. And these purification methods still form the basis of the water supply industry's strategy for delivering safe drinking water to the American public. The prevention of waterborne diseases is the cornerstone upon which even the most modern water treatment purification facility is built today.

But as industrial America has grown since the turn of this century, and as the synthetic chemicals industry has burgeoned since World War II, water purification has become less a problem of infectious diseases and more a problem of synthetic chemicals, particularly those that can lead to long-term adverse health effects, such as cancer. Unfortunately, the waterworks industry has been slow to respond to this changing tide, and the purification plants of nineteenth century design have gradually become anachronisms in twentieth century industrial America.

This was no better illustrated than when the Environmental Defense Fund and the Environmental Protection Agency released their respective reports on the cancer threats posed by industrial chemicals in New Orleans drinking water, calling attention to the inadequacies of the purification process in removing these chemicals. The treatment processes that have been so effective in combating waterborne diseases were found to be no match for carbon tetrachloride, benzene, dieldrin, and the numerous other contaminants that are ubiquitous pollutants of the nation's drinking water supplies.

As a consequence of these reports, the United States Congress enacted legislation, the Safe Drinking Water Act of 1974, which provided for the first time in this century authority for the federal government to establish and enforce national standards that would require modernization of public water supply systems. Unfortunately, as has been too often the case with our federal regulatory programs, change has been slow in coming. The U.S. Environmental Protection Agency has adopted the old Public Health Service Drinking Water Standards as national standards and has developed regulations to reduce the hazards created by the chlorination process, which has been found to contribute its share of carcinogens to drinking water. But no requirements have yet been promulgated that would upgrade water treatment practice to remove the numerous industrial and agricultural chemicals that are ubiquitous pollutants of most surface water supplies, or to correct the problems associated with water that leaches dangerous metals, such as cadmium and lead, from home plumbing systems.

Thus, citizens have been forced to seek home remedies, including bottled water and home water treatment devices in hopes of providing safe drinking water for their families. Approximately 20 percent of our population must depend upon private wells, and thus has no alternative but to become its own problem solver. But the problems often defy easy correction, and the citizen often must become his own chemist and engineer to effectively address his problem.

Although a flurry of government studies over the past decade has alerted citizens to the numerous drinking water hazards that may imperil their families' health, the public has had few places to turn for advice on the significance of local drinking water quality problems and remedies for them. And herein lies the value of this book. For Rodale Press has, as it has done countless numbers of times in other problem areas, provided the public with the first comprehensive analysis of drinking water problems and their solutions. Armed with this book, the citizen will have an important leg up in assuring that his family is provided with safe drinking water, however unresponsive federal and local agencies are to his problems.

Robert H. Harris, Ph.D.

Acknowledgments

This book represents the efforts of many people in many disciplines. The unifying factor among us is our concern about the safety of our drinking water, and its effect on the quality of our lives.

Above all, I would like to recognize Patricia M. Nesbitt, Environmental Consultant, of Strasburg, Virginia, for her major contribution to this book. It is her clear vision of the relationship between humans and water—even more than the volumes of data she supplied—that framed this book. Her contribution is best represented in the third chapter, describing the nature of water on this planet, but her work exists throughout.

The importance of water to our health and environment was brought into focus, then enlarged and enhanced, by Robert H. Harris, Ph.D., formerly Associate Director of the Toxic Chemicals Program of the Environmental Defense Fund, presently Member of the President's Council on Environmental Quality, who reviewed this book. His understanding of the incredibly grave water problems facing us today and in the future gave perspective to all working on this project when we most needed it.

Special thanks are due to several members of the Rodale staff: Joseph P. Senft, Ph.D., Research Scientist on the staff of the Organic Gardening and Farming Research Center at the Rodale Farm, and Thomas Stoneback, Ph.D., Project Director for Research and Develop-

x

ment at Rodale Press, provided technical assistance throughout the writing of *Water Fit to Drink.*

Emrika Padus, Managing Editor of *Prevention* Health Books, contributed her expertise on trace minerals to the book's fifth chapter.

Jean Polak, Editorial Coordinator, supplied the glossary of water terminology as an appendix to this book.

Very special thanks go to Carol Baldwin, Research Chief for *Prevention* Magazines and Health Books, and Eileen Mazer, Research Associate, who attempted to verify every word on every page, footnoting as they went. What resulted is a bibliography that offers additional reading for those who want to learn more about water.

News items, government reports, and personal insights were provided by Steve Franklin, a reporter and environmental writer for the Philadelphia *Evening Bulletin.*

Thanks, too, to Shirley DeEsch, who transformed worked-over copy into crisply typed pages, and to Lee Goldman, Executive Editor of *Organic Gardening* ® magazine, for allowing us time away from our magazine duties to complete this project.

The manuscript was also reviewed by my husband, Bill, and my sons Brad and Jeff, who, by asking so many good questions about our own water supply, made this book more complete for everyone.

Not a Drop to Drink 1

As Americans, we have always been a little smug about our water. The first question we ask a travel agent before setting foot on a foreign shore is, "Is the water safe to drink?" We kid about how the French are forced to drink wine because their water is so bad, and, fearing Montezuma's revenge, we order up Lomotil tablets before boarding the plane to Mexico.

But we have little reason to be smug. Our water is fast becoming a national scandal. Congress is grappling with the problem of how to protect, or at least compensate, people victimized by industrial poisons —buried decades ago—that have leached into local water supplies. The citizens of New Orleans are drinking water that has been used and reused by cities all along the Mississippi—a river called the colon of America—and suffering the consequences. Every city of any size has tested its drinking water and many found it wanting in purity.

It may come as a surprise to many, but our drinking water was never really pure. When the Lenape Indians dipped their gourds into a forest stream, they were downstream from *something*. Their water contained insects, bits of animal and human waste, rotting leaves, and all sorts of natural debris. As the nation was settled and grew more populated, people and their water developed an intimate relationship —often too intimate! The bacteria from human waste found its way into waterways and wells. By mid-nineteenth century, the United

States had suffered two cholera epidemics,[1] and typhoid was one of the ten leading causes of death.[2]

Yet, with typical American efficiency, the scourging bacterial diseases of cholera and typhoid fever, along with many other waterborne diseases, were controlled by the use of chlorine. As early as 1908, chlorine had been added to the municipal water supply for Jersey City, New Jersey.

At that time, Jersey City received its water from a private company that maintained a reservoir in rural and mountainous Boonton, New Jersey. The water supplied to the city was untreated and unfiltered, but generally clean and potable. However, Jersey City residents noticed that at times their water smelled and tasted bad. Some became ill. The city, which had a contract stating that all drinking water supplied to it would be wholesome, demanded that the private company honor the contract.

The water company considered the problem. The reservoir from which the water came was made by damming a small local river, the Rockaway. At certain times of the year, sewage from the towns along the river flowed unhindered into the reservoir. The company decided against building a filtration plant because of the expense. Instead, it chose to disinfect the water with chlorine.

What happened was quite dramatic. The bacterial count in the water dropped, and safe water was achieved at a cost much lower than with any other treatment devised at that time. Effective and cheap, chlorine not only killed germs, it also improved the taste and color of the water.[3]

It was the beginning of a new era. As the practice of chlorination became more widespread, the incidence of waterborne disease declined. By 1930, typhoid had dropped from among the ten leading causes of death in the United States to the twenty-sixth.[4]

In recent years, most Americans have assumed that, if the water in their cities and towns was less than pure, chlorine had taken care of the problem before it reached the kitchen tap. Yet, water is not clean and not necessarily safe to drink. While there is relatively little danger of contracting typhoid or cholera, there is very real danger of swallowing mercury, lead, arsenic, vinyl chloride, chloroform, carbon tetrachloride, pesticides, and more—depending on where you live and where your water comes from. Some of these pollutants are known to cause cancer in laboratory animals.

Cancer is so frightening, most people thrust the thought of it from their minds—quickly. There is an unspoken feeling that people who get sick—especially from cancer—deserve it. It's not a nasty thought, just a common one.

"Cancer of the lung? Oh well, I always knew that he smoked too much."

"Heart attack, you say. Well, she was really carrying a lot of weight for a long time."

It is as though heart disease and cancer, like syphilis, are earned through wanton, unclean, self-destructive behavior. But sometimes the things that cause disease are insidiously innocent. Like drinking water.

To celebrate my birthday, my husband took me and our two sons to a fine restaurant in downtown Philadelphia. From our windowside table, we could watch the Schuylkill River meander to the oil refineries and eventually out to sea. The late afternoon sun glinted on the heavy silverware and caught the stream of water the waiter poured into our goblets. The scene was one of serenity and familial peace until our younger son reached for his glass of water.

"Don't drink that!" my husband barked. "Put that glass down."

As heads turned all around us, my husband sheepishly explained to us that a recent newspaper account reported that Philadelphia's water contained asbestos particles, along with a bouillabaisse of 155 compounds, including at least 7 known to cause cancer.[5]

"Look, boys, this stuff could kill you," he said. And the kids were treated to Shirley Temples.

But in day-to-day living, Shirley Temples are hardly a viable alternative to water. Every time we make soup, mix up frozen orange juice, fill the coffee pot, or boil a pot of spaghetti, we are introducing pollutants to our bodies. If we don't want nicotine in our systems, we can refrain from smoking. If we want to avoid nitrates, we can do without ham or bacon. But we cannot do without water.

Once, a fakir in India survived for eighty-one days with absolutely no nourishment. But even he could not have survived for more than five days without water. It is unique to our survival because it is not only an essential nutrient, but also *the* major component of the body. Anywhere from 57 to 70 percent of what we are is made of water— with the leaner among us having the higher percentage, since fat is without a high water content.

Our bodies are made up of 100 trillion cells, each bathed in fluid. Moreover, each individual cell is filled with fluid. Fluid travels the 60,000 miles of veins and arteries within us, it lubricates our joints and soft tissues, and fills all the little hollows in our bodies. Water is the basis of our blood and our tears.

Each day our lungs exhale one-third of a quart of water, and our skin sweats away about a pint through the 2 million sweat glands in our skin. We excrete one-and-a-half quarts of water a day. All this must be replaced because the balance of water within the body is precise. If what goes out and what goes in do not come within 1 to 2 percent of each other, we feel either thirst or pain. The hypothalamus, a small section of the brain above the spinal cord, controls the feeling of thirst. If a body loses 5 percent of its total water, that person will suffer hallucinations. A loss of 15 percent will be fatal.[6]

Too much is as bad as too little. Remember that the intake and loss must balance. Forcing water was an ancient torture, causing the victim to experience nausea, weakness, mental confusion, tremors, convulsions, coma, and, finally, death.

Water is in many ways the key of life. We can't live without it.

———————

John Sarnicky and his wife had a happy life. In 1970 they settled in Pennsylvania's beautiful Lehigh Valley. He got a good job as a machinist at the local Mack Truck plant, and they bought a home in the town of Ormrod because John thought it would be a great place to raise a family.

They were settled and secure in their lives, and awaiting the birth of a child when the local health officials knocked on the door, telling them not to drink the water, cook with the water, or even bathe in it. The water had been poisoned with trichloroethylene (TCE), a chemical from industrial wastes, in concentrations about 3,000 times the recommended health standard.

The health officials could not tell the Sarnickys what the long-term effects of drinking the water might be, but they said TCE was suspected of causing cancer. However, a local newspaper carried an article saying that TCE can affect a pregnant woman's placenta.

"When you find out that your kid was poisoned before he was even born—that hurts," said John.

TCE is known to cause liver and kidney cancer in laboratory animals. Nobody is sure if it is a carcinogen for people, and nobody knows if the chemical can be passed through the mother's body into the fetus. For the Sarnicky family, it was a time of anxious waiting.

The water in Ormrod was poisoned by a landfill operation which dumped chemicals into abandoned iron ore mine shafts about 1,200 feet from the town well. According to records from the State Department of Environmental Resources, Western Electric Company of Allentown, the closest city of any size to Ormrod, had dumped about 3,000 gallons of wastes a week at the site between 1968 and 1969. Between 25 and 50 percent of those wastes were trichloroethylene.

State health officials investigated the landfill in 1971, noting that the groundwater from the landfill flowed directly toward the town well. It was not until August of 1976, however, that residents were told their water was unsafe. Health officials said their years of inactivity—years when the town was drinking poisoned water—stemmed from a lack of knowledge about the nature of TCE.[7]

A new water supply was found for Ormrod the following year. In the future years, the townspeople will have to continue to watch and wait—wait to see if they get cancer.

Our Water Is Chemical Soup

There probably isn't one single drop of naturally pure water anywhere on the face of the earth. And there never has been. Water is an excellent solvent, managing to pick up particles of rock and organic bits as it travels through the earth; as surface water, all manner of things fall or get dumped into it. Scientists group these many materials into several categories.

One category consists of microbiological contaminants, including some varieties of bacteria, viruses, and parasites that cause waterborne diseases and bad tastes. Health officials frequently test treated water for coliform, an "indicator" bacteria, to make certain they fall within standards of acceptability. The usual treatment to eliminate microbes is chlorination, which has been enormously effective. Viruses, and certain kinds of bacterial spores and cysts, are more resistant to chlorine than are the coliform bacteria. In fact, viruses may not always be effectively removed. On occasion, outbreaks of waterborne diseases do occur—usually because of a breakdown in the chlorination procedure or because of deficiencies in the system, such as broken pipes that allow bacteria to enter water after it has been treated.

Although chlorine works well as a disinfectant, its side effects are arousing considerable alarm. This additive is now known to interact with other elements in water to form carcinogenic compounds. The problem caused by chlorination will be discussed in Chapter 2.

Solids suspended in water make up the second large category of water contaminants. Little bits of clay, oil, and mineral fibers may be harmless by themselves but can serve as transportation for other pollutants which might be released later within the body. But some of these suspended solids are far from harmless—like asbestos. Some of the highest counts of asbestos in water have been found near large cities and industrial centers. Many studies have been done on the effects of asbestos on the human body, and it has been determined that the inhalation of asbestos can cause asbestosis and some forms of cancer.[8] Recent studies on laboratory animals and preliminary evidence from an epidemiologic study of gastrointestinal cancer and its relationship to asbestos in drinking water in the San Francisco area has increased concern that swallowing asbestos as well as inhaling it can cause cancer.

The third category of contaminants is comprised of inorganic solutes, which are mostly trace minerals. A report called "Drinking Water and Health" done in 1977 by the National Research Council Committee on Safe Drinking Water listed sixteen metals: barium, cadmium, chromium, lead, mercury, silver, beryllium, cobalt, copper, magnesium, manganese, molybdenum, nickel, tin, vanadium, and zinc.

This report said that evaluating these elements and their effect on the human body is a tricky business. For one thing, some of these trace metals in small amounts can be quite healthful (like zinc), while others in small quantities (like lead) can be deadly. What's more, many of these metals which are good for you in small doses are killers in large amounts. To further complicate matters, adverse health effects may be caused by a combined intake from water, food, and air. If the quantities of an element taken in food and air are unknown, there's no saying how much can be safely drunk in water.[9]

Nevertheless, these elements, from barium to zinc, have been in water a long time—long enough to have been studied, and for most to have maximum acceptable levels set for them. Their relationship to human health has been pretty clearly defined. But the next category of contaminants in water is quite another story. Called "organics," they sound harmless enough, suggesting some association with organic gardening and health foods. Some organics, however, are the antithesis of health food, and the nemesis of pure water.

Organic Compounds

The word organic has a specific meaning in chemistry, and describes any compound based on carbon. Carbon dioxide, methane, and butane, as well as fat, starch, sugar, and protein are examples of natural organic compounds. In addition to compounds found in nature, modern science has created synthetic organic compounds. Nylon, for example, is a synthetic made from compounds found in oats. Most synthetic fibers and plastic products are made from natural raw materials, and are thus called synthetic organic compounds.

Drinking water contains both natural and synthetic organic compounds, and they can create health problems. Some organics are the things that nightmares are made of.

The Hudson River Valley in New York State is a place of great scenic beauty. It is the only fjord in the United States. The land rises in palisades on one side of the river, affording magnificent views. The valley is a land of legend, of Rip Van Winkle and the Headless Horseman. The Sunday sailors who take their Snarks and snipes for a run in the summer breeze can look into the water and see fish swimming in large schools. In fact, the number of fish seems to be increasing each year. Sea fish, too, swim up the Hudson to lay their eggs in fresh water and then return to the Atlantic. Indeed, the Hudson looks as if it has defied the odds and remained pristine in spite of the chemicals dumped by riverside industries. But looks are deceptive.

The Hudson River is so polluted with chemicals that the river's fishing industry was shut down by the state in February 1976. Toxins such as benzene, polychlorinated biphenyls (PCBs), tetrahydrofuran, toluene, and chloroform (among others) were found in such high quantities that fish caught in the Hudson are considered unfit to eat.[10]

A 1977 study by the Environmental Protection Agency identified 280 organic compounds in the waterways. A joint study by two environmental activist groups, the New York Public Interest Research Group (NYPIRG) and the Environmental Defense Fund (EDF), found that some of the most potent cancer-causing toxins and chemicals known to cause birth defects were present in the water that served the 150,000 people in and around Poughkeepsie. And, if the plans set forth by the Army Corps of Engineers are accepted, this water will also serve 10 million additional people in the New York metropolitan area.

Unfortunately, the pollution of the Hudson River is not an isolated incident, and the residents of Poughkeepsie are not exceptional as victims of their drinking water. They share a common problem with the people of New Orleans, whose drinking water has been identified by the EPA as having sixty-six organic chemicals, some of them toxic. The EPA said there would be fifty fewer cancer deaths among white males each year if the cancer-causing agents were removed from the water.

Other guests at the chemical feast include:

• Philadelphians, who drink water from the Delaware River and are subjected to varying amounts of PCBs, a dangerous toxin.[11]

• The former residents of Love Canal, a small housing development built on a filled-in canal near Niagara Falls. Their drinking water was contaminated by a landfill containing eighty-two chemical compounds, including eleven known or suspected carcinogens.[12]

• Some residents of Canton, Connecticut, who have been advised to buy bottled water because seven of the town's wells contain extremely high levels of chemicals known to cause cancer in laboratory animals.[13]

• The taxpayers of Provincetown, Massachusetts, who had to cough up $460,000 to secure a new water source because theirs was threatened with gasoline contamination.[14]

- People who ate the shellfish from Chesapeake Bay after Kepone was discharged into those waters.[15]

Such problems exist across North America. In 1975, the EPA found the drinking water of seventy-nine cities polluted with traces of organic chemicals. At that time, the agency's administrator Russell E. Train said, "People should not react with any sense of panic, but they should know there is a problem."

Organic compounds come from five basic areas: natural sources; runoff, spills, and accidents; industrial discharge; sewage treatment plants; and water treatment.

The natural sources include leaves that fall on waterways and break down as they float along, waste products of aquatic plants and animals, organic acids from decomposition of organic matter in soil carried off the land when it rains or snows, and domestic wastes. Generally, as organic matter decomposes, humus and the associated humic acids are produced. There is no way to prevent these humic acids from entering water supplies. They do not cause health problems by themselves, but toxic materials such as pesticides and soil sterilants can combine with the humic acid to cause problems. And when chlorine is used to disinfect water, it interacts with humic acid to form chloro-organics, some of which are known carcinogens.

These chloro-organics are the second source of organics in drinking water. The most frequently identified of these compounds is known as trihalomethanes (THMs), of which chloroform is usually the most prevalent. Chloroform was once used as an anesthetic, but was banned by the Food and Drug Administration in 1976 because of its carcinogenic properties. Other THMs are known to cause mutations in laboratory animals, and are highly suspected of being cancer-causing agents in people. It is unfortunate that chlorination, which was thought to disinfect water cheaply and effectively, has turned out to be a double-edged sword.

The third and fourth sources of organic compounds are direct and indirect dumping of wastes into our supply of water. They are called point and nonpoint sources of pollution. When a contaminant can be traced to an individual source—for example, a pipe discharging chemicals from an industry into a river—it is a point source. A nonpoint source means the contamination is diffuse, such as a pesticide that runs off large masses of land and drains into a stream or seeps into a well.

Industrial waste is thought to be a major contributor of synthetic organics to drinking water. Considering the nature of the chemical processing industry, it is no wonder that toxic chemicals have made their way into our kitchen sinks. An estimated 500 to 1,000 new chemicals are introduced into our nation's industry each year, adding to the more than 10,000 chemical compounds that already exist.

Certainly, with all these new chemicals finding their way into our lives and bodies, someone must be keeping track, making judgments about safety, and perhaps keeping the most dangerous ones off the market. But no. Industrial disclosure laws, the patent process, and the customary privacy surrounding research and development programs make monitoring and surveillance nearly impossible.

For instance, it wasn't until great numbers of Michigan livestock and poultry began dying mysteriously that people knew even that the culprits—polybrominated biphenyls—were being made commercially, let alone being distributed. In fact, most compounds that aren't intended for human consumption aren't tested for toxicity to any significant degree. What happens then is that toxicity isn't discovered until it has done some harm, possibly years or even decades after it was put into use. Such is the case with vinyl chloride used to make polyvinyl chloride (PVC). In industry and consumer products for almost thirty years before it was found to cause cancer, it has been recently discovered that PVC pipe, a staple in the plumbing industry, will contaminate water with vinyl chloride if the water stands in the pipe for any length of time.

The other major point source of pollution is the effluent, often chlorinated, that comes from waste-water treatment plants. Generally, the effluent contains a good bit of organic matter of human origin, some trace elements, and organic and inorganic compounds kindly donated by industrial sewers. Chlorination of this effluent is one source of chlorinated compounds in waterways and in our drinking water. Chlorinated compounds are also found in waterways used by power plants. The water withdrawn for cooling the boiler and processing steam is first disinfected so that no microbial growth clogs up the internal plumbing. Power plants use about half the amount of chlorine that waste treatment plants do.

The fourth major source of organic compounds in drinking water is nonpoint sources, particularly farmland. The intensive application of pesticides and fertilizers has an adverse effect on water quality. In fact, the Illinois Pollution Control Board has considered a measure unprecedented in United States agriculture—state regulations that would limit the amount of fertilizer a farmer can spread about.[16] In central California, wells that supply fresh water to much of the area have suffered a sharp increase in nitrate levels. Fertilizer is the suspected cause.

What does all this mean to you? If your water contains nitrates, you are facing two health hazards. The first is a disease called methemoglobinemia—where the nitrate affects the blood so that it no longer carries oxygen through the body. Untreated, it leads to death. This disease most frequently affects infants, who are commonly called "blue babies" when they contract the disease. Healthy adults can consume greater quantities of nitrates in drinking water. In fact, many state health departments set two acceptable nitrate levels: below 45 milligrams/liter for infants, and at 45 milligrams/liter for others.[17,18]

The second risk is that the nitrates will change in the body, becoming nitrosamines, some of which can be carcinogenic. In fact, some preliminary studies have suggested a link between high concentrations of nitrates in drinking water and cancer of the stomach—but no firm conclusions have been made yet.[19]

Radioactivity in Water

Possibly one of the more dramatic hazards found in drinking water is posed by radioactivity. A Grade B, 1950-style motion picture could be made, called *The Creature Who Drank Radium-228.* It could star Steve McQueen, recreating the role he played in *The Blob.* But we would all be in the cast—because we all drink radium-228. And radium-226. And strontium-90. And many other atomic age goodies.

It is unfair to blame all the radiation in our drinking water on nuclear development: nature provides cosmic rays and radioactive elements in the earth's crust and atmosphere. This natural radiation, called background radiation, is unavoidable.

Of course, as the use of atomic energy becomes more prevalent, we potentially may be exposed to more and more radiation. Already the development and use of atomic energy for weapons, power plants, and medical techniques have spread radioactive substances throughout the world. There are spills and accidental releases of radioactive substances, and fallout from bomb tests. There is leaching of radioactive wastes from mines and milling operations, and from plants where nuclear products are fabricated.

Broomfield, Colorado, has a 40-acre drinking water reservoir that's lined with, of all things, plutonium. Residents of this Denver suburb recently learned their drinking water is radioactive.

According to the county health director, Carl J. Johnson, M.D., the beta radiation in the local water exceeds both EPA and state limits by almost 100 percent. Dr. Johnson also discovered the water is contaminated with plutonium, a substance considered ten to fifteen times more dangerous than other radioactive materials, and one which can remain in the environment for more than 500,000 years.

He says the danger from plutonium is even greater than that posed by excessive alpha and beta radiation, even if plutonium is present at levels now considered safe by the EPA. Moreover, he cites a study showing that chlorination of drinking water can magnify that danger by more than 1,500 times.

He explains: "The Environmental Protection Agency drinking water regulations provide for a limit of 5 picocuries per liter for radium 226 and radium 228 combined. Both are bone seekers and alpha emitters. However, plutonium is considered by most experts to be fifteen times more prone to cause cancer and other health effects than radium. When this ratio of biological effectiveness is considered, a level of 3.03 picocuries per liter of plutonium—a level found in one composite sample—may be considered equivalent to 45 picocuries per liter [hereafter abbreviated pCi/l] of radium, a very excessive figure. Moreover, some experts believe that plutonium is not fifteen times but two hundred times more hazardous than is radium."

The radiation, according to Dr. Johnson, stems from the effluent released by the Schwarzwalder uranium mine into a local creek that leads to the Long Lakes reservoir, serving the North Table Mountain Water District.

The danger to the public is compounded by two factors. One is that chlorination acts on plutonium, changing it into a more soluble form that is readily absorbed by the intestinal tract. Dr. Johnson cites a study showing that this more soluble plutonium results in concentrations in the bone and liver of animals 1,570 times greater than concentrations resulting from the less soluble form that exists prior to chlorination. The second factor is that plutonium is more easily absorbed when present in water than when present in food. Frequently phytates and phosphates in food "bind up" plutonium, as well as other metals, making it less available for digestion.

The danger of radiation contamination may extend beyond the boundaries of the North Table Mountain Water District. An environmental impact study has shown levels of plutonium as high as 2.5 pCi/l in water taken from the Arapahoe Aquifer, the source for many water wells east of the city.[20]

In northwestern New Mexico, high levels of radium have been discovered in some water samples drawn from wells near uranium mines and mills. The source of this contamination is waste water from these mines, which frequently is stored in lagoons. As time passes, the water evaporates, leaving behind radioactive material that can seep into groundwater supplies.

According to an EPA report released in 1975, seepage introduced 0.41 curies of radium into the shallow aquifer in Bluewater Valley. Further sampling near the Jackpile mine, outside Paguate, New Mexico, showed levels of radium-226 ranging from 0.31 to 3.7 pCi/l. Of all seventy-one samples collected, the EPA found six to have more radium-226 than the 3 pCi/l allowed by the Public Health Standards.[21]

Both background and newfound radiation have affected drinking water. But even the best scientists can't tell us specifically what effect radioactivity will have on our health and what amount of radiation is tolerable in water—if any. Minute traces of radioactive material are found in almost all drinking water. The concentration and composition of these radioactive particles vary from place to place, depending primarily on the radiochemical composition of the soil and rocks through which the water has passed.[22]

In a midwestern area that encompasses parts of Iowa, Illinois, Wisconsin, and Missouri, the natural levels of radium-226 and radium-228 are significantly higher than the national drinking water standard of 5 pCi/l. About 1 million people in this area drink well water that contains anywhere from 3 to 80 pCi/l of radium-226. These concentrations of radium pose a risk of bone cancer for the folks who drink this water.[23]

Exposure to radiation is, indeed, harmful to health and should be avoided whenever possible. People have always been exposed to background radiation. However, as the environment is filled with more radioactive material, our species will be under a great deal more stress from radiation than it has ever known.

The people who are the most sensitive to radiation are the young —the fetuses, infants, and children—because of their rapid cell division and the fact that radiation affects cell division. The effects of exposure to radiation can be divided into three categories—increased incidence of tumors, developmental abnormalities, and death.[24]

There are standards set for the amount of radionuclides allowed in water, but these figures do not guarantee anyone's health safety. The Environmental Protection Agency states that, at the doses allowed, "the bodily effects are not expected to be necessarily nonexistent but rather nondetectable."

The act of drinking water is so ordinary that most people do it without thought. However, when someone becomes aware of the dangers that lurk in the tap, the coffee pot, the soup kettle, or the pitcher of iced tea, the ordinary act of drinking water becomes tinged with uneasiness.

NOTES

1. *Encyclopedia Brittanica,* vol. 5 (Chicago: William Benton, 1961), p. 616.

2. Patrick R. Dugan, "Use and Misuse of Chlorination for the Protection of Public Water Supplies and the Treatment of Wastewater," *American Society of Microbiology News,* March, 1978, pp. 97–98.

3. *Drinking Water and Health* (Washington, DC: National Academy of Sciences, 1977), pp. 4–6.

4. Ibid., p. 98.

5. Warren Froelich, "Chemicals in U.S. Drinking Water," *Philadelphia Evening Bulletin,* March 10, 1978.

6. Luna B. Leopold, Kenneth S. Davis, and the Editors of Time-Life Books, *Water* (New York: Time-Life Books, 1966).

7. Warren Froelich, "Pa. Town Can't Use Its Water," *Philadelphia Evening Bulletin,* March 10, 1978.

8. *Drinking Water and Health,* pp. 161–168.

9. Ibid., pp. 302–303, 316.

10. Richard Severo, "Two-Year Study of Hudson River Finds Toxic and Cancer-Causing Chemicals That Threaten 150,000 Upstate," *New York Times,* September 29, 1977.

11. Warren Froelich, "Toxic Chemical Is Found in Fish from Delaware River," *Philadelphia Evening Bulletin,* April 25, 1976.

12. Donald G. McNeil, Jr., "Upstate Waste Site May Endanger Lives: Abandoned Dump in Niagara Falls Leaks Possible Carcinogens," *New York Times,* August 2, 1978.

13. Diane Henry, "Connecticut Moves on Polluted Wells: Defunct Chemical Company Called Culprit in Canton, But Another Is Ordered to Do Cleanup," *New York Times,* December 3, 1978.

14. "Massachusetts Town Is Seeking Funds to Save Its Water Supply," *New York Times,* November 19, 1978.

15. "Kepone Threat Halts Lower Bay Fishing," *Washington Post*, September 20, 1977.

16. George C. Bubolz, "Our Water Supply—How Safe Is It?: Farmers Must Stop Using Nitrogen," *Natural Food & Farming*, December, 1978, p. 1.

17. *Drinking Water and Health*, p. 418.

18. David Zwick and Marcy Benstock, *Water Wasteland* (New York: Grossman, 1971), p. 14.

19. S. R. Tannenbaum, M. Weisman, and D. Fett, "The Effect of Nitrate Intake on Nitrite Formation in Human Saliva," *Food and Cosmetics Toxicology*, vol. 14, 1976, pp. 549–552.

20. Personal communication with Carl J. Johnson, M.D., Director, Jefferson County Health Department, Lakewood, CO, October, 1979.

21. *Impact of Uranium Mining and Milling on Water Quality in the Grants Mineral Belts, New Mexico* (Region VI, Dallas, TX: U.S. Environmental Protection Agency, 1975), pp. 2, 3, 5.

22. *Drinking Water and Health*, p. 858.

23. Ibid., p. 860.

24. Ibid., p. 872.

Government in Action— Or Government Inaction 2

A lot of people dismiss the threat of dangerous drinking water. They've been drinking it all their lives and never caught anything from it yet. Most people believe that, if anything were *really* wrong with the water, the government would take steps to solve the problem.

Let's look at the first statement. If the water is bad, why aren't more people sick from it? First of all, a lot of people *are* sick from it —but researchers are just now looking at water as a prime suspect in causing illness. One reason drinking water has long gone unsuspected is that its ill effects may take decades to surface, decades of ingesting minute traces of something harmful each and every day. Meanwhile, the water looks, smells, and tastes okay: few could guess that drinking tap water is a potentially dangerous habit.

The federal government is aware of our troubled water. The Environmental Protection Agency (EPA) monitors pollutants in our environment and sets acceptable limits; it develops methods to clean the environment; and it enforces the laws and standards it has developed. Additionally, Congress frames legislation to control pollutants in the environment, and occasionally Congress and the EPA work together—Congress stating a broad goal for environmental cleanliness and asking the EPA to come up with the methodology to make it so, and to play enforcer. And that is exactly how the Safe Drinking Water Act was created.

Let's go back to 1974 and to New Orleans, the home of the Mardi Gras, the French Quarter, the Super Dome, and some really rank drinking water, and watch how the government operated. Residents of New Orleans were accustomed to drinking water that had a strange oily feel on the tongue. Sometimes it was yellowish. Sometimes it smelled. But, after all, New Orleans is at the mouth of the nation's largest sewer pipe, the Mississippi. Residents played the games of guessing where the source of today's water befoulment was. A leaky barge in Cairo? Sewage overflow in St. Charles?

But scientists didn't have to guess what was making the water so bad. Back in 1972, the Federal Water Pollution Control Administration looked into the Crescent City's drinking water (both raw and treated) and identified forty-six organic compounds. Incredibly, they said these compounds represented only 2 percent of the total chemical mess in the water. Based on this report, the EPA announced that the city's water was dangerous, especially to the very young and old, and to the ill.[1] The report was received with yawns and ho-hums from everyone except other environmentalists, who became worried about the quality of drinking water, not only in New Orleans, but all over the country.

At the same time the report was getting rave ho-hums, a bill called the Safe Drinking Water Act was lost in Congress. It had been shuffled about for more than two years, and it would be yet another two years before it would see the light of day.

Then, in 1974, the Environmental Defense Fund (EDF), a Washington-based environmental group, released a study that showed a correlation between cancer and the drinking water in New Orleans.[2] The EDF examined the average cancer mortality rates of all the parishes (counties) in Louisiana from 1950 to 1969. After taking into account the many variables generally associated with human cancer—urbanization, family income, occupation, source of drinking water—the EDF found that white males who were drinking water from the Mississippi River were considerably more likely to die of cancer than those who used groundwater, which was less polluted than the river water. The very next day, the EPA published a survey that showed there were numerous chemicals, some of them carcinogens, present in the New Orleans water supply.[3]

When the news reached the capitol, Congress was appalled. In just a little more than a month, they found the lost Safe Drinking

Water Act, and passed it into law. At the time of its passage, Congress mandated the EPA to find out exactly what was in the nation's drinking water, and to set limits for the harmful elements. Moreover, Congress asked the EPA to enforce those limits in waterworks throughout the country.

The story should have a happy ending, but it doesn't. As the EPA set about drawing up guidelines for safe water and setting maximum acceptable levels of contaminants, it became clear that the commonly found group of organics that had been identified at that time—called trihalomethanes (THMs)—was caused by the water purification process itself. When water is chlorinated, the chlorine interacts with humic acid and other natural debris in water to produce THMs. The most common THM is chloroform, and it is a carcinogen.[4]

In order to remove THMs from drinking water, treatment plants can take several actions. They can find a substitute for cheap, old-reliable chlorine, aerate water to dissipate it, or add chlorine to water later in the treatment process, when fewer or no humic acids are present to combine with chlorine, forming THMs. This last method was tried in Cincinnati, where the THMs were reduced by more than half simply by chlorinating when the water was cleaner, and by using less chlorine. The net result was not only a reduction in THMs but also a reduction in the cost of treating water.

Another way to rid water of THMs, and other organic chemicals, is to filter them out with granular activated carbon (GAC). Waterworks managers, engineers, city managers, economists, and others launched a fierce fight to stop the regulation of organics because it required the use of GAC beds. Federal funds were not available, and the cost of the new system and other improvements would be carried directly by the water utilities, and, ultimately, the consumer.[5]

Eventually, the EPA knuckled under to the waterworks association and coalition of waterworks companies, withdrawing its guidelines for organics in drinking water. As a result, a good portion of the American public would continue to drink its water with chloroform.[6]

In the time the EPA spent dragging its feet proposing guidelines for THMs, the National Cancer Institute and the National Academy of Sciences announced that water is dangerous because of organic compounds. Ultimately, the EPA was sued by the Environmental Defense Fund to act more quickly. Finally in February of 1978, it proposed amendments to the "Interim Primary Drinking Water Regulations," setting a ceiling for specified trihalomethanes, and to require treatment for removal of other potentially hazardous synthetic organic chemicals. Larger water companies (those serving more than 75,000 people) could not allow the THMs to exceed 0.10 milligram per liter, or 100 parts per billion. These companies were given five years to comply with the guidelines.[7]

The EPA said, "The THMs are formed at the drinking water treatment plant when chlorine is added to kill disease-causing bacteria. They are an unwanted by-product of chlorine's reaction with naturally occurring organic substances in the untreated water. Most public water systems use chlorine to disinfect water, and *THMs are present in virtually every supply EPA tested* [author's emphasis]. THMs are the most common organics found in drinking water and they are also present in higher concentrations than other organics. EPA is concerned about chloroform because it has caused cancer in test animals and may pose the same risk to humans."[8]

When the guidelines were issued, the head of the EPA, Douglas M. Costle, said, "We know that some of these chemicals are dangerous. The lifetime exposure of our population to these chemicals poses a serious threat to public health. We are especially concerned about the increase in cancer risk."[9]

All this occurred in early 1978. By mid-1978 President Jimmy Carter addressed the Congress with a water policy message. He said his new policy was to achieve four basic objectives: improved planning and management of federal water programs; a new national emphasis on water conservation; enhanced cooperation between the federal government and the states on water policy and planning; and increased attention to environmental quality.[10]

The General Accounting Office of the United States (GAO) was asked to review the new policy and make comments. The GAO shot back with a sharp critique, noting that the policy made scant mention of the importance of water quality.[11] In its November 1978 report, the GAO said, "The drinking water supplied to most American homes today is generally considered good; however, a 1970 study indicates that water supply quality may be deteriorating. The high standards set by United States public water supply systems produced a steady decline in the number of outbreaks of waterborne diseases and illnesses. But that decline stopped in 1951, and there are indications that it may have begun to rise. In addition, the nation's water supplies are threatened by the careless use of hundreds of chemical compounds and the heedless disposal of toxic wastes."[12]

The GAO report went on to discuss the importance of creating new policies that complement the Safe Drinking Water Act. Moreover, it discussed the failure of the Clean Water Act of 1977 to set reasonable standards for the nation's waterways. This act required that the nation's streams be suitable for fishing and swimming by 1983.

The GAO said, "One of the real concerns with maintaining water quality is the increasing degree of water pollution that is being caused by nonpoint sources of pollution. [Note: Nonpoint pollution includes runoffs from agriculture and forest lands, mining and construction sites, and urban areas.] We testified in July 1978 that at the rate funds are being authorized for nonpoint pollution, it will be impossible for many of the Nation's streams to meet the 1983 goal of being fishable/swimmable. . . . If the 1983 goals are too costly for the Nation to obtain, the administration needs to address what the national priorities will be and what quality of water the Nation can realistically obtain under funding and staffing constraints."[13]

Later in the report, the GAO noted that according to one study the cleanup of just urban runoff would cost about $199 billion. "Although it is obvious that controlling nonpoint sources of pollution will cost billions, only $600 million has been authorized under the Clean Water Act of 1977 to assist owners of rural property to install the best management practices for long-term soil conservation to improve water quality by reducing runoff,"[14] the report said. It concluded by saying, *"Thus, the Nation's water supply is not, in many cases, of sufficient quality to be used for drinking purposes* [author's emphasis]."[15]

Beneath all this bureaucratic wrangling, and attempted blame-placing, the quality of the drinking water remains, at best, the same. Valuable time was lost during the 70s—the decade that saw the infamous Love Canal disaster.

Unfortunately, these government reports and dramatic disasters produce in the American public a kind of tunnel vision. It is easy to believe that bad water is the exception—relegated to the Mississippi, to highly industrialized areas, to *someplace else.* Unfortunately, that is not true.

The EPA found water problems in 246 river basins, covering

almost the entire nation.[16] And if your water comes from a well rather than a river, ponder this. The EPA also listed 32,254 disposal sites with potential hazardous waste problems.[17] These are dumping sites for suspected or known toxic chemicals that can leach into the underground water supply to poison drinking water.

The problem of polluted drinking water is everywhere. It is not a matter, simply, of legislation, treatment, reports, and bureaucratic fumbling. It is a matter of life and health.

While the government attempts to regulate and clean our water, the illegal dumping, the leaching, the runoff continue. Some of the most dangerous elements in water are the organics—things like pesticides, and industrial wastes. Other organics are the by-products of chemicals intentionally put into our water: chlorine and fluoride.

Chlorine—A Closer Look

A research team from the Columbia University School of Public Health began looking into the deaths of housewives in seven New York State counties. As in any good mystery, there was a prime suspect responsible for the deaths of these good women.

The team, headed by Dr. Michael Alavanja, examined each female death from cancer of the gastrointestinal or urinary organs in the years 1968 to 1970. Because the prime suspect in these cancer deaths was chlorine, the researchers chose to study women: 85 percent of them were housewives, and they likely drank the same water throughout the day. Since the population in these counties was very stable, most of the women had spent their lives in the same locale, drinking from the same source, year in and year out.

The scientists found that women in the study who drank chlorinated water ran a 44 percent greater risk of dying from cancer of the gastrointestinal or urinary tract than those who drank unchlorinated water.

"To our knowledge, this is the first time a significant statistical relationship has been demonstrated between human gastrointestinal and urinary tract cancer mortality and chlorinated drinking water," Dr. Alavanja said.[18] He recently completed a second study on men with essentially the same statistical correlation between cancer and drinking water.

In addition to cancer, chlorinated water has also been linked to high blood pressure, and anemia. Studies in Russia have shown that men drinking water with 1.4 milligrams of chlorine per liter have higher blood pressure than those drinking water with only 0.3 or 0.4 milligram per liter.[19]

In this country, John Eaton, Ph.D., Professor of Medicine at the University of Minnesota, found that chlorine had a deleterious effect on red blood cells. He made this discovery in 1973 while studying patients who had developed severe anemia during treatment at two artificial kidney centers in Minneapolis—both centers used chlorinated water in their kidney dialysis machines. In his laboratory, Dr. Eaton found that these patients' red blood cells had been severely damaged by the chlorinated water. His findings were confirmed at a third artifi-cial kidney center in Minneapolis. This third center had chlorine-free water, and its patients did not develop anemia. Now, new federal water standards in many states prohibit exposing dialysis patients to chlorine above 0.1 part per million.[20]

Recent EPA studies indicate that the synthetic chemicals formed in the chlorination process are far more dangerous than chlorine itself. Called trihalomethanes (THMs), they are the largest group of synthetic chemicals found in drinking water. "They are found in virtually every drinking water supply that is disinfected with chlorine, and not uncommonly at concentrations of several hundred parts per billion," according to an EPA report.[21]

These "not uncommon" concentrations far exceed the limit proposed by the EPA of 100 parts per billion or 0.10 milligram per liter.[22]

Among THMs, chloroform is the most common. A recent EPA survey of treated drinking water indicated that 95 to 100 percent of the finished water they tested contained chloroform. The treated water in Miami, Florida, for example, surprised researchers with more than three times the recommended amount of chloroform.

When someone drinks water containing chloroform, the chemical is rapidly absorbed in body fat and tissues. Chloroform was once used in cough syrups and other drug and cosmetic products. The United States produced and consumed hundreds of millions of pounds of chloroform each year. Chloroform was banned July 29, 1976, for use as an anesthetic but is still used in chemical processing and in research.[23]

Tests performed on mice and rats demonstrated that chloroform causes cancer, most frequently as malignancies of the kidneys and liver.[24] The National Academy of Sciences concluded that "data strongly support the contention that chloroform is carcinogenic in at least one strain of mouse and one strain of rat. . . . It is suggested that strict criteria be applied when limits for chloroform in drinking water are established."[25]

A second common trihalomethane found in water as a result of the chlorination process is bromoform (tribromomethane). Of eighty water samples surveyed, it was present in about one-third. Another survey showed bromoform was present in water after chlorination, even when none was present in the raw water.

Bromoform is used commercially as a solvent for wax and oil, and as an ingredient in fire-resistant chemicals. Few animal studies have been made on the health effects of consuming bromoform—it was never intended for consumption.[26] However, because its chemical properties are similar to chloroform, its general effects may be similar.

Yet another child of chlorination is chlorobenzene. It showed up in nine out of ten water supplies surveyed by the EPA. While some chemicals can be defused by the body's metabolism, it appears that chlorobenzene becomes a more potent poison once inside the body, acting as a depressant of the central nervous system. This chemical is used in the manufacture of insecticides and dyes.[27] Although the effects of this chlorination by-product have never been studied in humans, it is very probable that such effects will be similar to those determined in animal studies.

The Other Chemicals

It is well past midnight in Philadelphia. Just 30 miles south of the spot where Washington threw a coin across the Delaware, a trucker sits pondering the same river. It is cold, and it is late. The trucker looks at the factories—they are brooding and empty, lining the old cobblestone streets. Even the late shift has gone home. He checks the rear view mirror. No traffic, no cops. All clear.

The driver quietly shifts into reverse, then very carefully backs his truck down a narrow street that runs between two warehouses and lets out on the river. Quietly—very, very quietly—he dumps a load of chromium waste into the river. In the darkness, no one sees the water turn red. He pulls out and is gone, barreling his truck north—away from the city and its strange-smelling water.[28]

On the very same night, only a few yards up river, police nab another trucker for illegally dumping ammonia into the Delaware. Sometimes truckers dump the chemical loads within feet of drinking water intake valves.[29] The midnight dumping is a serious problem that is becoming more common. As government regulations became more stringent, companies that used to dump rather freely were in a bind. But not for long. A new profession has sprung up in the last decade. It's called gypsy hauling, and these truckers will haul *anything* and dump it *anywhere*, for a price. No questions asked.

In Charles City, Iowa, a pharmaceutical manufacturer dumped chemicals into the Cedar River watershed, which provides the drinking water for at least 10 percent of Iowa's population. An EPA study showed twenty-four "priority pollutants" (EPA jargon for chemicals that cause cancer or hereditary defects). One of the chemicals, orthonitroaniline, has been found 65 miles downstream, in shallow wells that provide the drinking water for the city of Waterloo, Iowa.[30]

Preliminary studies show that residents of Charles City and Waterloo have elevated levels of bladder cancer compared with others living in similar areas. "We can't make a definite link between the two, but there is some reason for concern," said the executive director of the Iowa Department of Environmental Quality.[31]

In Niagara Falls, New York, at least three potentially dangerous dump sites were discovered. One dump, the Hyde Park landfill, contains—by its owner's own estimate—more than 80,000 tons of hazardous wastes. Some of those wastes run off in streams that pass surrounding homes and schools, and eventually flow into Lake Ontario. One resident near the Hyde Park dump says she has watched six or seven of her pet cats die. "They go down by the stream and it starts like eating the skin and the fur off their feet. Then they die a few days later," the woman said.

Two of the three dumps are dangerously near the city's water supply. One is located next to the water treatment plant, the other is near the Niagara River's banks, just upstream from the plant's intake pipes.

The major concern of federal and state officials is the presence of a chemical called trichlorophenol (TCP)—used to make other chemicals like 2,4,5-T and the Agent Orange herbicide. Worst of all, dioxin, one of the most poisonous of all substances known to man, is formed as a contaminant in the process of making TCP. According to a *New York Times* report, dioxin is "as virulent as botulism or shellfish toxin. . . . Three ounces minced small enough into New York City's water supply could wipe out the city."

The Hyde Park landfill apparently holds some 3,300 tons of trichlorophenol, which means it may contain a *ton* of dioxin, small amounts of which trickle into the stream that runs through the landfill.[32]

At the same time the existence of these chemicals was made public, some residents of the neighboring Love Canal were holding an ironic Christmas celebration. The remaining few people gathered in the office of the Love Canal Homeowners' Association to affix decorations to a small Christmas tree. The ornaments were made of plastic freezer container caps, and were attached to the tree with pretty red and green bows. Each ornament had a name written on it. But where you might expect to see "Suzie" or "Grandma Price," there were, instead, names like nitrobenzoyls, trichlorophenol, benzene, lindane, and the names of seventy-seven other compounds found in the air and water in this neighborhood.

These partygoers were some of the 550 residents who have not been able to evacuate their homes (unlike the 237 residents whose homes were bought by the state of New York for $10 million). The association president, Lois Gibbs, raised a glass of punch and toasted her home town—the boarded windows, the empty houses, the bulldozers, and the canal.[33]

When an entire town faces an extreme health threat and is bought out by the state for millions and millions of dollars, it is a page-one story in almost every newspaper in the country. Less dramatic cases are happening all around us, but rarely even make it to the back page. Lisa Wood, editor of a newsletter for the Clean Water Action Project in New Jersey, said, "Several incidents have occurred to endanger the safety of our drinking water supplies in Middlesex County. Often these episodes have received only limited coverage by local newspapers. Some of the contaminants are carcinogens or suspected carcinogens. Most of them have necessitated the closing of groundwater supplies."[34] The people of that county won't see it reported on the TV evening news, and will continue to drink their water, unaware of the dangers it poses to them.

In one episode in Middlesex County, New Jersey, a health officer saw a tanker truck spilling fluid as it traveled along an industrial road that is adjacent to the public drinking water well field. The truck parked within ten feet of one of the wells and continued its discharge. The health officer called local police, who arrested the driver and impounded the truck.

When the truck was examined, it was found to contain poly-chlorinated biphenyls (PCBs), a known cause of cancer in animals. This chemical then filtered down into the water wells, where traces were found for the following seven months. But no matter. The public was in no danger. These same wells had already been closed because of heavy metal concentrations.[35]

In other parts of New Jersey, six wells—including two public wells in Camden County and one in Middlesex—were found to contain levels of toxic chemicals that exceeded recommended standards for drinking water.[36]

In Canton, Connecticut, water wells were found to be con-taminated with industrial chemicals known to cause cancer in animals. These compounds appear to have leached through decaying valves buried by a chemical company that had been out of business for six years.[37]

The list of towns, cities, and farms that suffer from contaminants in water is endless—and not completely known by anyone. What all this means is that our health may be in danger without our knowledge.

Here's a startling example of how widespread the problem is. In an EPA water survey, eight of ten samples contained cyanogen chlo-ride. That's the stuff tear gas is made from. Scientists know that a large dose of this cyanide is poisonous, but they don't know what happens to someone who consumes small amounts daily.[38]

Another common water pollutant is carbon tetrachloride—clean-ing fluid—making an appearance in the water of eight out of ten municipal utilities tested. It's known that children can die from drink-ing small amounts of carbon tetrachloride. This chemical also causes kidney failure, heart failure, gastrointestinal upset, mental confusion, cirrhosis of the liver, and blood enzyme changes. It has been shown to cause cancer in laboratory animals, and cancer risks have been projected for the United States population: scientists estimate that a lifetime exposure to carbon tetrachloride in drinking water at 10 parts per billion "would produce one excess case of cancer for every 50,000 people exposed," or 4,400 deaths in this country.[39]

Both the World Health Organization and the National Cancer Institute have suggested that between 60 and 90 percent of cancers are caused by environmental contaminants—with maybe as much as 90 percent being chemical in origin.[40] The National Cancer Institute compiled a list of over 1,700 organic compounds found in water in both the United States and Europe. Of this number, less than 25 percent have been tested for toxicity.[41]

But common sense would lead us to believe that organics in our water are dangerous and cause cancer. The incidence of cancer has risen steadily since 1900, when it ranked as the eighth leading killer. Today it is second only to cardiovascular disease. This increase is much greater than an increase that could be attributed to the longer life expectancy in this country, or to the reduction in the number of deaths caused by other diseases.

More frightening than this present-day rise in cancer is the fact that cancer will probably become much more prevalent in the future: by its very nature cancer takes a long time to develop after exposure to carcinogens. The cancers we see today are the result of exposures that took place over the last twenty or thirty years. The increasing number of carcinogens and other toxic substances in the environment is likely to cause more cancer in the future.[42]

William Lijinsky, Ph.D., a toxicologist with the Frederick Cancer Research Center in Frederick, Maryland, told a 1977 conference on drinking water problems, "We are a bit arrogant about what we know about cancer. . . . So how can we dismiss any toxic substance as being a potential carcinogen? Some toxins magnify the effect of others and thereby greatly increase the total toxic burden."[43]

Fluoride, Added by Legislation

Many water supply systems routinely add fluoride to help protect dental health. From the first day fluoride was put into drinking water thirty years ago, a wild debate has raged.[44] Anti-fluoridationists charge

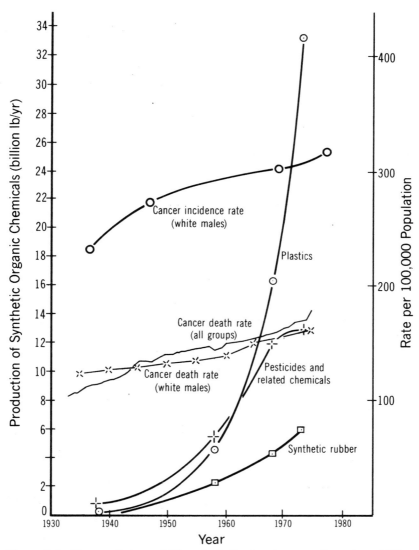

Figure 2-1: Cancer Rate and Chemicals Production as a Function of Time. *(Source: Correspondence with Robert Harris, Ph.D., member of the President's Council on Environmental Quality, Washington, DC.)*

As the production of chemicals grows, the rate of cancer deaths takes a sharp turn upward. It is important to note that cancer rates lag behind chemical exposures by thirty to forty years. The post-1950 increase in the production of synthetic chemicals probably will not be reflected until after 1980.

that fluoride is a poison—sometimes used to kill rats—that should be kept out of water. But most dentists and public health officials believe that fluoride is of great benefit to the general public.

What exactly is this hotly disputed additive? Fluoride is a product of the element fluorine, a gas that is one of the most violently reactive substances known. Fluoride is found in the earth's crust, in sea water, and in the human body as a trace element.

In proper doses, fluoride helps prevent tooth decay. In very high doses, it damages the teeth by mottling them, and can do harm to bones and overall health. On the other hand, some studies have shown that high doses of fluoride can restore certain kinds of hearing loss,[45] help build bones in cases of osteoporosis[46] (a bone disease), and even ward off heart attacks.[47] Most of these claims are still just speculations, but the one claim that holds true is that in proper doses fluoride helps prevent tooth decay.

Research on the role of fluoride in tooth decay prevention began in the 1930s, when residents of a large number of cities with naturally fluoridated water were compared with residents of cities where little or no fluoride was present. These studies led the Public Health Service to conclude that a level of fluoride at 1 part per million could lead to a major reduction in tooth decay.

A later fluoride study done in the 1940s compared the dental health of teenagers in Grand Rapids, Michigan, with that of teenagers in Aurora, Illinois. Grand Rapids had a fairly low level of natural fluoride in its drinking water—0.15 part per million—while Aurora, Illinois, had a high level of 1.2 parts per million. As might be expected, the kids in Aurora had better teeth than those in Grand Rapids. The Aurora kids had an average of five decayed teeth, while the Grand Rapids kids had thirteen.

Then, the drinking water in Grand Rapids was fluoridated with 1 part per million. At the end of the test period, a new batch of sixteen-year-olds was examined and found to have an average of only seven decayed teeth—the problem was cut by almost half. Other tests performed over the years have consistently shown that fluoride added to drinking water cuts cavities by as much as 65 percent.

Here's how fluoride works. Teeth are made of a mineral called apatite, a complex calcium compound. When fluoride is present in the diet in just the right amount (1 ppm), it becomes incorporated into the

Table 2-1: Possible Effects of Fluoride in the Diet

Fluoride Dose[a]	Possible Result
1.0–2.5 mg F/day	Level of total F[b] intake in a child in a community with fluorided water. This is the amount *recommended* for prevention of a large fraction of tooth decay.[1]
0.5–1.0 mg F/day	Amount of fluoride *in addition to* the recommended amount that is needed to produce an unacceptable risk of dental fluorosis.[2]
1.5–3.5 mg F/day	A level of intake in children that is unacceptably high, because of possible dental fluorosis.[3]
2.0–5.0 mg F/day	Average amount ingested by an average adult in a community with fluoridated water.[4]
5 or more mg F/day	Amounts that may produce changes in the skeleton that may or may not be "harmful."[5]
20–80 mg F/day	Doses that, over 10 years or more of exposure, will produce clearly evident, irreversible, serious damage to the bones, in healthy adults.[6]
50–200 mg F/day	Doses that are used in therapy for certain diseases of the bones.[7]

Sources:

1. Food and Nutrition Board, National Research Council, *Recommended Dietary Allowances* (Washington, DC: National Academy of Sciences, 1979).

2. R. Aasenden and T. C. Peebles, "Effects of Fluoride Supplementation from Birth on Human Deciduous and Permanent Teeth," *Archives of Oral Biology*, vol. 19, 1974, pp. 321–326.

3. This estimate is simply the sum of the first two dose levels.

4. L. Kramer et al., "Dietary Fluoride in Different Areas of the United States," *American Journal of Clinical Nutrition*, vol. 27, 1974, pp. 590–594.

5. A. Singh and S. S. Jolly, "Chronic Toxic Effects on the Skeletal System," World Health Organization, Monograph No. 59, *Fluorides and Human Health* (Geneva: World Health Organization, 1970), pp. 238–249.

6. Committee on Medical and Biologic Effects of Atmospheric Pollutants, National Research Council, *Fluorides* (Washington, DC: National Academy of Sciences, 1971).

7. D. Rose and J. R. Marier, *Environmental Fluoride—1977* (Ottawa: National Research Council of Canada, 1978).

a. mg F/day = milligrams of fluoride daily.

b. F = fluorine.

teeth, making them harder and more resistant to decay. Scientists believe the critical period is during pregnancy, when the fetus is forming its baby teeth, up to about seven years of age.

Since almost everyone agrees that fluoride is beneficial to dental health, there would seem to be no room left for debate. However, some scientists speculate that the long-term effects of fluoride may not be totally beneficial. For one thing, they point out that a "safe" level of 1 part per million of fluoride in water does not determine how much a person will consume. Many substances contain goodly amounts of natural fluoride—like ordinary tea. And if one drinks several mugs of steaming orange pekoe every day, each made with fluoridated water, the acceptable level of fluoride may be surpassed.

Other scientists believe that long-term exposure to fluoride may cause cancer. Recently a Pennsylvania judge ordered a water company

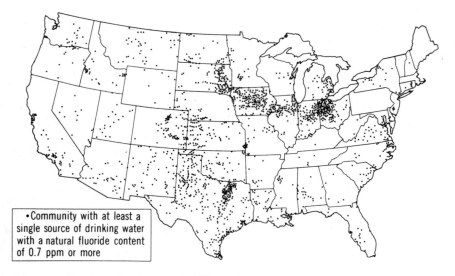

• Community with at least a single source of drinking water with a natural fluoride content of 0.7 ppm or more

Figure 2-2: Natural Fluoride in Water Supplies.
(Adapted from Water Atlas of the United States, *by James L. Geraghty et al. [Port Washington, NY: Water Information Center, 1973], plate 47, by permission of the publisher.)*

Each dot on this map represents a community having at least one water source that contains 0.7 ppm (parts per million) or more of natural fluoride. Approximately 8.4 million people reside in the communities shown on the map: 40 percent of these people live in Texas, 52 percent live in Illinois, Indiana, Iowa, Ohio, and South Dakota.

that serves twenty-seven communities to stop adding fluoride to the water because residents of those same communities had an elevated number of cancer deaths. His decision was based on a study done by Drs. John Yiamouyiannis and Dean Burk of the National Health Foundation, who claimed that fluoride could cause cancer.[48] His decision was later overturned by a higher court. Fluoride was also withdrawn from the water supplies of Jersey City, New Jersey,[49] and Eugene, Oregon.[50] There are no absolute findings that say fluoride causes cancer, just as none says it does not. But the stakes are big. About 105 million people in the United States drink fluoridated water every day.[51]

A lot of the pro and con arguments about the addition of fluoride to water are political as well as medical. Many people believe strongly that adding a prescription drug to the general water supply is unethical

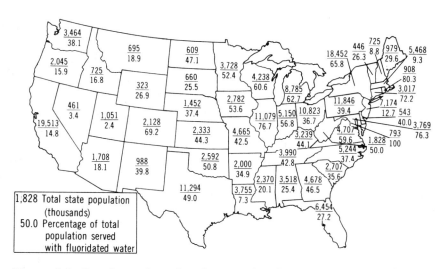

Figure 2-3: Population Served with Natural and Added Fluoridated Water.
(Adapted from Water Atlas of the United States, *by James L. Geraghty et al.* [Port Washington, NY: Water Information Center, 1973], plate 48, by permission of the publisher.)

The map shows the total population of each state and the percentage of that population served with fluoridated water. Note that every state in the country has fluoridated communities. In Maryland, Wisconsin, Virginia, Illinois, and Connecticut, over 90 percent of the population on public water supplies has access to fluoridated water. Furthermore, Connecticut, Minnesota, Illinois, Delaware, Michigan, South Dakota, and Ohio have statewide fluoridation laws.

because it leaves individuals without a choice. Those who don't want to drink fluoride are stuck. Another point is that fluoride does very little good for teeth after childhood. In effect, an entire population takes a drug when it benefits only those under seven years old. A final argument: the whole notion of putting a drug into the water supply smacks of Big Brother to many, and some people worry that tranquilizers or contraceptives may be the next drugs down the pipe.

While we know fluoride helps reduce tooth decay when given to children before and during the formation of their permanent teeth, its long-term effects are still a matter for debate. In areas of the country where water is naturally high in fluoride, people apparently are not troubled. But then, it took three-quarters of a century to discover that our ally chlorine has a potential carcinogenic effect when combined with substances found naturally in water.

And so it would seem prudent to take drinking water "straight" —with no additives, wherever and whenever possible. As for children's dental health, there are several alternative ways of getting fluoride's protection.

Mouthwash is one answer, as learned in a three-year study by the National Institute of Dental Research. More than 70,000 school children in seventeen communities without fluoridated water were given fluoride mouthwash to use once a week. Their tooth decay was cut by about one-third. The director of the institute estimates that 20 million children could participate in a school-run mouthwash program.[52]

Another option is fluoride tablets, which could be administered to school children each day. One study showed that kids who chewed a fluoride tablet each day had a significant reduction in the number of fillings they needed. The study continued for six school years—or until the kids were in the first year of junior high school. The cost of a year's supply of tablets came to just over $2 per child.[53] Another study showed that chewable fluoride vitamin supplements given to children eighteen to thirty-nine months old reduced their number of cavities by about 37 percent in the first six months, 55 percent after one year, and 63 percent after two years.[54] In Israel, children are given their fluoride in fruit juice: the juice apparently enhances the uptake of fluoride by three times.[55]

So, it is possible to safely omit fluoride from our drinking water, and thereby avoid the worry of long-term exposure. Where no public health or community fluoride treatment program is available, most dentists will give patients a prescription for fluoride tablets.

Environmentalists vs. Industry: The Long, Long Tale of How Duluth Finally Got Asbestos-Free Drinking Water

On June 15, 1973, the federal government warned parents in Duluth, Minnesota, not to allow their children to drink the water. The Environmental Protection Agency had found the supply full of asbestos fibers, and feared that those who drank it would risk getting gastrointestinal cancer. The EPA's Dr. Larry Plumlee limited the warning to children because adults had been drinking the water so long that "stopping now would not do much good."[a]

The EPA said the asbestos fibers probably came from the wastes dumped by the Reserve Mining Company, 60 miles away in Silver Bay, Minnesota. This company daily dumped 67,000 tons of taconite tailings into Lake Superior, which supplies drinking water to Duluth and other lakeside towns.[b]

In 1969 the state of Minnesota tried to get Reserve to stop using the lake as a dump, but it was not until February 17, 1972, that the United States Justice Department filed suit against Reserve.[c] Sixteen months later, the citizens of Duluth were warned that the water was too dangerous to drink.

The ensuing trial between the United States and Reserve weighed the economic impact of closing down the 3,200-employee company against the possible danger to 150,000 drinkers of asbestos-polluted water.[d,e] United States District Court Judge Miles W. Lord took a stand against Reserve. He said, "It has been clearly established in this case that Reserve's discharge creates a serious health hazard to the people exposed to it. The exact scope of this potential health hazard is impossible to accurately quantify at this time. Significant increases in diseases associated with asbestos exposure do not develop until 15 to 20 years after the initial exposure to fibers. . . .

"At present the court is faced with a situation where a commercial industry is daily exposing thousands of people to substantial quantities of a known human carcinogen. . . . Under no circumstances will the court allow the people of Duluth to be continuously and indefinitely exposed to a known human carcinogen in order that the people in Silver Bay can continue working at their jobs."[f]

Fourteen months after the trial began, Lord ordered Reserve mining to shut down its operation until it could find a way to dump its tailings on land, the way other taconite companies do.

At that time, Reserve was making a profit of $60,000 a day. Each year the plant was in operation, there was a 90 percent return to its owners, the Armco and Republic Steel companies. In other words, for each dollar Armco and Republic initially invested in Reserve, they got back ninety cents each year Reserve continued in operation. Because their losses would be staggering, the two steel companies immediately turned to the Eighth Circuit Court of Appeals.[g]

The Reserve Mining Company operation was shut down a total of forty-seven hours when Lord's order was reversed. Not only did the appeals court overrule Lord, but it also dismissed him for being "greatly biased."[h] The Eighth Circuit Court ordered Reserve and the state of Minnesota to find and agree on a safe land disposal site. In the meantime, Reserve continued dumping asbestos laden taconite tailings into Lake Superior.[i]

By May 1976, the case of the United States of America vs. the Reserve Mining Company had entered its third phase. Federal District Chief Judge Edward J. Devitt, who had replaced Lord, fined Reserve $1 million for violating the state water discharge permits from May 20, 1973, to April 20, 1974.[j] The company was still under orders to find a land dump site that met with the state's approval.

In July of 1976, the negotiations between the mining company and the state had fallen apart. The company wanted to dump tailings on a plot quite near the lake shore. The state objected to the site because it feared asbestos fibers would be washed into the lake. The two parties were at an impasse. Judge Devitt, hoping to press them into some sort of agreement, ordered the company to phase out dumping into the lake within one year, by July 7, 1977.

Reserve again appealed to the Eighth Circuit Court, but this time the appeals court upheld the district court: Reserve would have to stop using Lake Superior for disposal by mid-1977. However, the court suggested that, if circumstances should change, Reserve could have an extension.[k]

In April 1977, Reserve's proposed land dump site—the one close to Lake Superior—was accepted by a weary state. The court gave Reserve three years to switch its dumping operation from the lake to the new land site.[l]

From the time the EPA first suspected that Reserve was responsible for the asbestos in Duluth's drinking water, 1968, until the company was supposed to cease dumping in Lake Superior, twelve years had elapsed. In the interim, residents drank water the city imported from other towns, and ultimately installed filters to remove the fibers from Lake Superior water.[m] Only time will reveal the health damage suffered by the citizens of Duluth.

a. Stuart Auerbach, "Cancer-Causing Asbestos Is Found in Duluth Water," *Washington Post*, June 16, 1973.

b. Robert H. Harris, Edward M. Brecher, and eds., "Is the Water Safe to Drink: Part 1: The Problem," *Consumer Reports*, June, 1974, p. 6.

c. George C. Wilson, "Environmental Backfire: Winners Seen Losers in Mining Case," *Washington Post*, July 8, 1974.

d. George C. Wilson, "Health vs. Dollar Loss: Key Issue Rides on Reserve Mining Ruling," *Washington Post*, May 25, 1974.

e. Harris, *Consumer Reports*, p. 6.

f. Wilson, *Washington Post*, May 25, 1974.

g. Wilson, *Washington Post*, July 8, 1974.

h. "Mining Firm Fined $1 Million for Waste," *Washington Post,* May 5, 1976.

i. Harris, *Consumer Reports,* p. 6.

j. "Mining Firm Fined $1 Million Dollars," *Washington Post,* May 5, 1976.

k. "Reserve Mining Gets Deadline on Pollutant," *Washington Post,* October 29, 1976.

l. "Settling at Silver Bay," *Washington Post,* April 16, 1977.

m. Wilson, *Washington Post,* July 8, 1974.

NOTES

1. Lawrence Wright, "Troubled Waters," *New Times,* May 13, 1977, p. 35.

2. Ibid., p. 35.

3. Ibid., p. 36.

4. *Drinking Water and Health* (Washington, DC: National Academy of Sciences, 1977), p. 179.

5. Leslie Major, "Watering Down the Standards," *Environmental Action,* September 23, 1978, p. 14.

6. Wright, *New Times,* p. 38.

7. Environmental Protection Agency, "Interim Primary Drinking Water Regulations: Control of Organic Chemical Contaminants in Drinking Water," *Federal Register,* February 9, 1978, Part II, p. 5756.

8. Environmental Protection Agency, "EPA Proposes Controls on Suspected Carcinogens in Drinking Water," *Environmental News,* January 25, 1978, p. 2.

9. Ibid., p. 3.

10. General Accounting Office, "Review of the President's June 6, 1978, Water Policy Message," *Report by the Comptroller General of the United States,* November 6, 1978, p. 2.

11. Ibid., p. 7.

12. Ibid., p. 29.

13. Ibid., p. 30.

14. Ibid., p. 31.

15. Ibid., p. 34.

16. Nancy L. Aldrich, ed., "Slants & Trends," *Clean Water Report,* November 21, 1978, p. 226.

17. Nancy L. Aldrich, ed., "Slants & Trends," *Clean Water Report,* December 5, 1978, p. 237.

18. Michael Alavanja, Inge Goldstein, and Mervyn Susser, *Report of Case Control Study of Cancer Deaths in Four Selected New York Counties in Relation to Drinking Water Chlorination,* Columbia University School of Public Health, December, 1976. Personal communication with Michael Alavanja, D.P.H., Assistant Professor in Environmental Health and Epidemiology, Hunter College, New York, NY, July, 1977.

19. A. I. Bokina, V. K. Fadeeva, and E. M. Vikhrova, "29875q State of the Cardiovascular System of Men After Long-Term Use of Chlorinated Drinking Water," *Chemical Abstracts,* vol. 77, 1972, p. 29875.

20. Personal communication with John Eaton, Ph.D., Professor of Medicine, University of Minnesota, Minneapolis, MN, October, 1979.

21. Environmental Protection Agency, *Federal Register,* p. 5759.

22. Environmental Protection Agency, *Environmental News,* p. 2.

23. *Drinking Water and Health,* p. 713.

24. Ibid., p. 714.

25. Ibid., p. 695.

26. Ibid., p. 695.

27. Ibid., pp. 709, 710.

28. Stephen Franklin, "Bootleg Waste Dumping: Ecological Time Bomb," *Philadelphia Evening Bulletin,* May 28, 1978.

29. Ibid.

30. Nancy L. Aldrich, ed., "Slants & Trends," *Clean Water Report,* October 24, 1978, p. 210.

31. Ibid., p. 210.

32. Donald G. McNeil, Jr., "Three Chemical Sites Near Love Canal Possible Hazard: New Niagara Falls Dumps Reported Even Larger," *New York Times,* December 27, 1978. Personal

communication with John Iannotti of the New York Department of Environmental Conservation, Bureau of Hazardous Wastes, Albany, NY, October, 1979. Personal communication with Walter Hang, Molecular Biologist with the New York Public Interest Research Group, New York, NY, October, 1979.

33. "A Joyless Noel for Niagarans Still Remaining," *New York Times,* December 27, 1978. Personal communication with Lois Gibbs, president, Love Canal Home Owners' Association, Niagara, NY, October, 1979.

34. New Jersey Public Interest Research Group, "Threats to Safe Drinking Water in Middlesex County," *Clean Water Action Project Newsletter,* May, 1978.

35. Ibid.

36. "Most Jersey Water Meets Standards," *New York Times,* December 27, 1978.

37. Matthew L. Wald, "Tests Widened on Water Found Tainted in Canton," *New York Times,* November 22, 1978.

38. *Drinking Water and Health,* pp. 717, 719.

39. Ibid., p. 703.

40. New York Public Interest Research Group and Environmental Defense Fund, *Testimony of Walter Hang Before the Poughkeepsie Common Council Regarding the Subject of Carcinogens in Drinking Water and Health,* December 5, 1977.

41. H. F. Kraybill, C. Tucker Helmes, and Caroline C. Sigman, "Biomedical Aspects of Biorefractories in Water," *Proceedings of the Second International Symposium on Aquatic Pollutants,* September 26–28, 1977, p. 431. Personal communication with Herman Kraybill, Ph.D., Scientific Coordinator for Environmental Cancer, National Cancer Institute, Bethesda, MD, October, 1979.

42. New York Public Interest Research Group, *Testimony of Walter Hang Before Poughkeepsie Common Council Regarding the Subject of Carcinogens in Drinking Water and Health.*

43. Personal communication with William Lijinsky, Ph.D., toxicologist with the Frederick Cancer Research Center, Frederick, MD, October, 1979.

44. *Drinking Water and Health,* p. 370.

45. "Sodium Fluoride Shows Promise for Otosclerosis," *Modern Medicine,* June 10, 1974, p. 160.

46. "Osteoporosis," *Medical Letter,* November 5, 1976, p. 99.

47. "Habits, Fluoride Aid Heart?," *Science News,* April 15, 1978, p. 230.

48. Lawrence Walsh, "W. View Fluoride Halted: Cancer Question Cited by Judge," *Pittsburgh Press,* November 16, 1978.

49. "Jersey City Council Irks Pros," *Fluoridation News,* July–September, 1978.

50. "Fluoride for Sale After Oregon Voters Say 'No,'" *Los Angeles Times,* July 6, 1977.

51. "Dental Care Should Be Initiated by Physicians," *Modern Medicine,* April 15, 1979, p. 24.

52. "Schools Urged to Use Fluoride Mouthwash to Cut Tooth Decay," *New York Times,* July 19, 1978.

53. William S. Driscoll, Stanley B. Heifetz, and David C. Korts, "Effect of Chewable Fluoride Tablets on Dental Caries in Schoolchildren: Results After Six Years of Use," *Journal of the American Dental Association,* November, 1978, pp. 820–824.

54. "Dental Caries Reduced by Chewable Fluoride," *Modern Medicine,* August 21, 1972, p. 69.

55. "Israeli Children Receive Fluoridated Fruit Juice," *Medical Tribune,* February 25, 1973, p. 3.

The Nature of Water 3

Barbara Coyle is a pleasant woman, recently awkward with the roundness that late pregnancy brings. Her husband, Matt, has a carefully cultivated nonchalance, but tonight he is anxious. His black eyebrows knit over the bridge of his nose as he looks out the living room window at the Schuylkill River, some 50 feet from the front of his house, across a narrow road.

"People are moving their cars to higher ground," Matt says. "The river looks like it might flow over the road."

"Oh, don't worry about it. It couldn't come up as far as the houses," Barbara replies. With her feet propped on the coffee table, she wiggles her toes. She views them complacently, feeling secure in the snugness of her home. The drumming of the rain on the roof is a sound that reminds her of Girl Scout camp-outs. But Matt is threatened by the intensity of this January rain. She is much more frightened by the recent ice and snow storms. "Rain never hurts anything," she thinks to herself.

She coaxes Matt away from the window and into the kitchen for a cup of cocoa. And as they sit talking, the insistent hammering subsides to splashing sounds on the windowsill, and they both know the worst is over.

But as they ready for bed, they are interrupted by their neighbor pounding on their back door. The neighbor is also a friend, with whom they share "twin" homes that are attached along the center wall.

Floyd apologizes for the interruption, and tells them not to drink the water. "It's gone bad. You can see it," he explains. "This happened once before, when the river flooded. And I know how to fix it. But for now don't drink it, unless you boil it first."

Barbara and Matt trust Floyd, who has been a surrogate uncle to them since they moved to Reading, Pennsylvania. So they boil a big pot of water before they go to sleep, allowing them to have their usual breakfast in the morning. But Barb and Matt were both confused about why the well they share with Floyd had gone bad.

The well is located in their backyard, which is banked quite high above the river. "How could the water get into our well when it didn't get up to the house?" she asked Matt. He wasn't sure. But somehow the polluted river water had mingled with the water in their well, and, for now, the water was no good to drink.

During the next week, Barbara sent a water sample to a local laboratory for testing. Within a day she received a form saying that her water had been examined by the membrane filter method and was found to contain 550 coliforms per 100 milliliters of sample. Except for the box where the word "polluted" was marked with an X, the form meant nothing to her.

"How bad could the water *be?*" she asked a co-worker. "After all, it isn't fecal coliform. And it doesn't really look bad anymore."

In deference to her unborn baby's health, Barb Coyle refrained from drinking the water. But she did brush her teeth with it, once accidentally made a batch of frozen orange juice with it (which she eventually boiled, then threw away). She no longer boiled her water for twenty minutes, but only brought it to a boil.

She was mighty surprised to learn, several weeks after the flooding episode, that the bacteria count of 550 was 550 times greater than the accepted norm, and that fecal coliform was not mentioned in the laboratory report because the water was not tested for it.

Suddenly frightened, Barbara decided to learn how her well became polluted, and what she and Matt would have to do to clean it up.

The Mighty Molecule

Two atoms of hydrogen and one atom of oxygen combine to form a molecule of water. But this is no ordinary molecule! Should anyone try to separate these three atoms, they're in for a tough time. They remain intact, whether frozen solid or heated until they become vapor. The reason these atoms stick together so well is that they have a strong bond based on the way the electrons within the atoms combine to form the water molecule. Each atom has a shell around its nucleus, formed by electrons. The hydrogen atom has a shell with one electron, but it has space for two. The oxygen atom has six electrons, but space for eight. When the hydrogen and oxygen atoms come together, the "empty" spaces in each are filled by the other.

Once these electrons have combined, they form an extremely stable molecule with a rather lopsided shape. The hydrogen atoms sit at one end of the molecule, which is positively charged and attracted to other substances that are negatively charged. The oxygen end of the water molecule is negatively charged and attracted to things with a positive charge. The molecule works a lot like a bar magnet, where one pole is positive, the other negative.

This electrical attraction allows water molecules to combine into a fluid that does some remarkable things. For example, the negative/oxygen end of a water molecule will draw to it the positive/hydrogen end of another water molecule. Ultimately, countless molecules form a tight network that is surprisingly strong. For instance, when water is poured into a wide pan, the surface tension is strong enough to float a small metal grid.

Water can climb because of this electrical attraction. Picture water molecules in a glass. The hydrogen end of the water molecule may be attracted to oxygen atoms. Those atoms on the surface of the water touching the glass will reach upward, stretching the surface of the water into a crescent shape. This motion is capillary action, and accounts in part for water's ability to climb from the roots beneath the soil to the top of the highest tree. Molecules move in a kind of hand-over-hand action, each pulling another along behind it.

And so these mighty molecules defy gravity. They are even more amazing when frozen into ice. The ends with the two hydrogen atoms are attracted to the ones with the one oxygen atom to form a structure like a pyramid, an ice structure that is "hollow in the center and takes up more space than liquid water." For this reason, ice is lighter than water, and in fact it floats. While the fact that ice floats does not seem very remarkable, just think what would happen if it *didn't.*

Instead of rivers forming a protective skin of ice in winter, allowing the water beneath it to flow, picture that ice dropping to the bottom of the riverbed. Without a protective skin, more and more water would freeze, and soon only a thin film of liquid skimming the ice would be available to us. Without the moderating influence of water —the ability of large bodies of water to retain heat even in winter— earth's climate would change radically. The temperature each day would drop and soar hundreds of degrees. The winds, no longer softening with moisture, would howl across the land, parching further the already dry soil. In short, if ice didn't float, the earth wouldn't exist as we know it.

Incidentally, almost all other substances shrink and become heavier when they are frozen. They sink. Ice defies all odds when it floats.

Finally, water molecules exist as vapor. Although water vapor is always invisible, it often condenses into droplets so that it can be seen as steam, mist, or massed in billowy clouds. And it is this vapor that allows the earth's constant water supply to circulate around and through this planet.

The Hydrological Cycle

There is no more or less water now than when the earth was formed. We have today basically the same water we started with, evaporated, condensed, recirculated, and reused over the eons. It covers three-quarters of the earth's surface, its molecules moving constantly between the sea, the air, and the land. The study of these independent movements of all forms of water (fluid, ice, or vapor) is known as hydrology. And the circulation of earth's water is called the hydrological cycle.

Essentially, the hydrological cycle views all water in a cyclic movement that never stops. The ocean's water continually evaporates as vapor into the atmosphere. Some of this water vapor will condense to form rain or snow. When this occurs over the ocean, the water is returned to the ocean directly. When precipitation falls on land, it may be caught by plants or intercepted by buildings and paved surfaces. But eventually, water reaches the ground, where it will go to one of three places. It could fall or flow into lakes, ponds, or puddles. Or, it might flow overland, gathering in depressions and rivulets to become surface runoff in the form of creeks, streams, and rivers. Some of the rain seeps into the ground to be stored as groundwater. In this manner, water

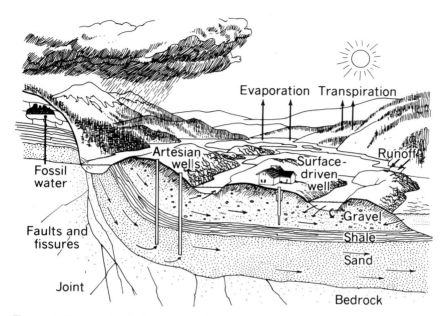

Figure 3-1: The Hydrological Cycle.

moves from the ocean to the land, feeding lakes, rivers, and groundwater. Ultimately, it travels back to the ocean, where it is again evaporated, condensed, and returned to the earth's surface.

The movement of water through the hydrological cycle is accomplished by the sun's heat, the winds, and the pull of gravity. Water is evaporated from the oceans and land by the heat of the sun, is carried as vapor in the air by the winds, and through gravity it falls as rain or snow and is pulled downhill, and sinks into the soil. As water moves through the cycle, a host of ingredients is added and then subtracted, affecting its drinkability. In some stages of the cycle, water is the best of beverages; in other stages, for example, as sea water, it is a poison to humans.

Oceans

Ocean water is rich in dissolved salts and minerals, in atmospheric gases, in both nutrients and poisons, and in sediment and suspended particles. Sea water is, in fact, a very complex solution of organic and inorganic salts generated over the millenia—the result of dissolved rocks, volcanic activity, and even a little meteoritic fallout from beyond earth's primal atmosphere. About three-quarters of sea salt is sodium chloride, the same stuff we shake on our burgers. The remainder is salts of magnesium, calcium, potassium, and other metallic ions. These salts render sea water undrinkable for man.

In recent years, the ocean has become a watery dump for all kinds of pollution, including oil spills, radioactive wastes, and the sludge from sewage treatment plants. In time, everything mankind makes ends up in the sea.

Precipitation

Water vapor moves through the atmosphere, eventually massing into clouds and condensing as rain, hail, sleet, and snow. The vapor (and eventually precipitated water) is affected by everything it touches in the atmosphere. While most naturally occurring gases are inert and do not react with water, some have a strong influence on the quality of water that falls to earth, even though they make up only a small percentage of the volume of air. One such gas is carbon dioxide, which forms carbonic acid when it comes into contact with water vapor. Carbonic acid is relatively weak, with a pH of 5.7 to 6.4. Its presence causes rainwater to be slightly acid. Acidic rain dissolves some materials in the soil and rocks, and the water becomes laden with minerals—it becomes "harder."

Other atmospheric gasses can also act to make rainwater acid. Gaseous components of air pollution—sulfur dioxide, hydrogen sulfide, some nitrogen oxides, and hydrogen chloride—all act on water, forming strong acids, like sulfuric, nitric, and hydrochloric acids. The rain created in this chemical hodgepodge can be more acidic than vinegar.[1,2] In fact, in rural areas all over New England, the acid rain has begun to take its toll on vegetation, aquatic life, and the stability of the forests.

The increased acidity of the rain speeds up the decomposition of mineral soils, and quickly weathers rocks, stones, and concrete. Plants and trees find nutrients required for their growth and survival stripped from the soil.

Massive fish kills in the Adirondack Mountains in New York and in industrial Sudbury, Ontario, show how devastating acid rain can be.[3] And in cities, people mourn the loss of outdoor statuary, and watch the suddenly rapid degradation of old buildings and monuments.

Surface Water

Surface water is simply water on the surface of the earth. It is oceans and puddles, mighty rivers and creeks you can jump across. The source of surface water is rain—along with snow, sleet, and hail. About half the population in this country drinks from streams, lakes, and rivers.

Picture a week in April when it has rained hard three or four days in a row. The little creeks swell and turn muddy, lakes crawl up their banks and snare the trunks of trees that are usually landlocked. Rivers roar and rampage and allow newspapers to print old photos of past floods. Obviously, the more water earth receives from the heavens, the more surface water is momentarily found on earth.

Of course, the converse is also true. In times of drought, streams and even rivers dry up, lakes sulk back to their centers, and the babbling brooks are strangled.

So much for quantity. The *quality* of surface water depends on the quality of rainwater itself, what it picks up as it flows along, the biological life in the water, the ways it is used, and contaminants that get pumped into it. Surface waters often start out as high quality waterways and end up as chemical soup.

The headwater of a river, its source, is a mysterious thing. Can something as mighty as the Mississippi really begin as a dinky run of water? And is it possible a river so foul could be clean where it starts? Indeed, both are true.

Rain and snow fall in the remote, often mountainous, places where rivers begin. Precipitation falls on forests and fields, and slowly seeps into the earth. As it moves through the humus, it is filtered, purified, and enriched with oxygen. Gravity pulls the water down to the crevices and cracks in the earth—to water's highway, the riverbed.

As water moves along a riverbed, it erodes rocks and gravel, it licks at its sandy base, and dissolves many rock minerals, which become part of it. In its rush to the ocean, a river can gouge out huge canyons and become loaded with salts and minerals.

But many rivers are gentle. These don't carve Grand Canyons, and generally don't carry a heavy burden of salts and minerals in their waters. Instead, they have a slow and intimate relationship with the vegetation along their banks and in their beds. As a result, they carry organic matter that enriches the water and encourages aquatic life.

These aquatic organisms improve surface waters. They are a sort of natural quality control mechanism.

These little plants and animals keep waters clean by metabolizing dissolving nutrients, eating floating solid particles, and cleaning the bottom sediment of decaying matter. In clean lakes and rivers, a wonderful balance exists. The plants respire oxygen, which the fish and invertebrate organisms use. In turn, these living creatures respire carbon dioxide, which the plants need.

Unfortunately, this beautiful relationship has been upset by people. Some lakes and rivers receive huge amounts of sewage, industrial wastes, and agricultural runoff that not only burden the natural assimilation processes but can also destroy them. The result is an excess of nutrients which stimulate the growth of algae and other aquatic plants. The algae grow and bloom, becoming so dense that they block out the sunlight from deeper parts of the lake or river. And when the algae die, the decomposition consumes huge amounts of oxygen. The fish and invertebrates can no longer live in such a place. The process is called eutrophication. When a river or lake eutrophies, it is said to be dying.[4]

But, by far, the most important factor in determining the quality of surface water is the way the water is used. Since the human race began, people have settled along the banks of rivers, streams, and lakes. There, they enjoy a reliable source of drinking water, food, and a flowing path to other places and other people. The Ganges, the Nile, the Thames, and the Amazon all share that common bit of history.

In our own country, many large cities have developed along the banks of our rivers and lakes. The populations of those cities share with their ancient ancestors the common uses of water. But if our ancestors fouled their water supply, they simply moved on. Today we no longer have that luxury. Moreover, modern man has found additional ways to use water—for industrial processing, for boiler cooling, for recreation, and to lure tourists. And our rivers and lakes have proven to be a convenient place to dump waste—human and industrial.

About half the pollutants rivers and lakes receive are direct discharges from storm sewers, municipal sewage treatment plants, and industries. The other half comes from untreated and uncontrolled runoff from parking lots, lawns, agricultural areas, and construction sites. These pollutants do not always deteriorate the quality of surface waters noticeably, but in some rivers and lakes of already dubious quality, this additional pollution may be just enough to make the difference between acceptable and unacceptable water.

Now, rainwater runoff does not sound like a significant problem in water quality. After all, even acid rain is only rain. But let's take a look at a rainy afternoon in Boston.

The Common looks like a merry mushroom garden, with red and yellow and blue and striped umbrellas bobbing in the rain. It is a soft autumn rain—not a hurricane or even a thunderstorm. In fact, a WBZ radio weatherman predicted only ⅛ inch would fall—barely enough to dampen the newly reseeded lawn near the band shell.

But underground, a maelstrom is taking place. Water is surging into storm sewers all over the city. One-eighth inch of rain falling on Boston's 29,440 acres amounts to 100 million gallons of waste water, all headed for the same two sewage treatment plants. The plants are unable to treat the deluge. The plants also are unable to treat the city's sewage. The volume of water is so overwhelming that the two treatment plants simply open their gates and discharge both rainwater runoff and raw sewage into Boston Bay.

Sometimes municipal waste is discharged into a body of water as a matter of course—not just in the instance of rain. Engineers and water pollution officials assume that a river or lake will assimilate all wastes by means of bacterial digestion, aeration, and dilution occurring naturally in the larger water body. "The solution to pollution is dilution," they say. And back in the days when there was little pollution and lots of clean water, maybe dilution was an answer to waste disposal. But that day is long past.

When the amount of waste dumped into a body of water exceeds the capacity of the water to treat it, or when the wastes are poisonous to life in the water, the waterways cannot rejuvenate themselves.

Lake Erie is exemplary for the study of surface water pollution. The municipal wastes of more than 9 billion people enter Lake Erie, either directly or through tributaries. The cities of Detroit, Toledo, Fort Wayne, Loraine, Sandusky, Cleveland, Akron, and Erie all discharge partially treated waste water into the Lake Erie basin. Even with some partial treatment, these wastes flowing into the lake are the equivalent of the raw wastes of 4.7 million people. The most harmful pollutants in this waste are nitrogen and phosphorus, which act as fertilizers for algae. The next most harmful are the bacteria passed from bathroom bowl to lake, rendering the water unfit for human contact. Third, are the toxic chemicals and persistent organic compounds that get dumped into the lake along with the municipal effluent.

Lake Erie and its tributaries also receive at least 360 direct discharges of industrial waste not routed through a public sewage treatment plant. Among the industries contributing to the demise of the lake are manufacturers of automobiles, steel, refined oil, rubber, chemicals, plating processes, and food processing. Until recently, fewer than half of these industries had adequate waste treatment facilities. In fact, the Cuyahoga River that runs through Cleveland occasionally catches on fire—its water is so overwhelmed with pollution.

Lake Erie still lives in the memories of some very old people— people who remember it when it was a lake. A Great Lake. A body of water pristine, cold, running with fish, and bordered with evergreens. A place where people could picnic, take a vacation, swim and boat.

Surprisingly, in the past few years, scientists studying the lake have found that reports of its death may have been greatly exaggerated. They found that, when pollution was reduced, the lake made some small attempts to cleanse itself. This is an encouraging reminder that, with common sense and careful management, even the most polluted of our waters stands a chance of being saved.

Groundwater

Like surface water, groundwater comes from the sky. When rain falls, the soil soaks up some water and holds it like a sponge. The remainder either runs off or sinks deeper. There it accumulates, filling the spaces around the rocks deep below ground.

Sometimes groundwater makes a pretty splashy show, as when it collects as lakes in underground caverns. The Mammoth Cave in Kentucky is a good example of an exotic body of groundwater. Tourists poke about the cave's spooky recesses, and in the beam of a flashlight they can see a species of blind fish swimming throughout.

But most groundwater is not nearly so interesting. Rather, it simply collects below the subsoil, contained by the rock, sand, or gravel that underlies the surface. It can seep into river and stream beds, supplying water to them even during times of grave drought. But most groundwater is generally pumped out of wells.

Sometimes, in periods of heavy rain, surface waters rise and seep into the soil on either side of the banks. The water table rises, and the waters of the aquifer mingle with newly intruded surface waters.

This was the cause of pollution in the Coyle well. When the Schuylkill River rose above its usual level, its waters entered the aquifer into which the Coyle's well was drilled.

An aquifer is a porous rock formation that holds groundwater. The term comes from two Latin words—*aqua*, water; and *ferre*, to bring. This formation can consist of sand and gravel, or a layer of sandstone or cavernous limestone. Or, it could be a large mass of fractured rock with sizable openings. An aquifer can cover a few square yards or several thousand square miles. Some are but a few feet deep, while others span several thousand feet. At the bottom of an aquifer is a layer of bedrock that makes the formation fairly watertight.

When a well is dug, it taps the water contained in an aquifer. Whether or not a particular aquifer is good as a drinking water source depends on several of its physical characteristics. The kind of rock, sand, or gravel that form it will have a bearing on the purity and taste of the water it holds. The number of pores and cracks in the aquifer will determine how much water it can hold. The size and relationship of the openings within an aquifer also influence how fast water can flow through it. Aquifers composed of rocks that have openings large enough for water to flow freely are good to tap, because water can be taken out in useful amounts.

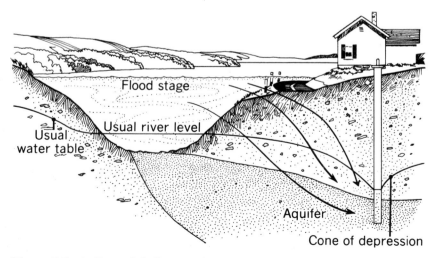

Figure 3-2: A Case of Pollution after Flooding.

Water exists in the ground in many forms, and is held in many places besides aquifers. There are two distinctly separate layers of water in the ground, separated from each other by the water table. The upper portions of the soil hold some water, where it is available to plant life. This damp soil is called the zone of aeration, where the spaces between the soil and subsoils hold air. Immediately beneath that layer is the zone of water saturation which saturates all pore spaces between the soil or subsoil particles. The line between these two layers is called the water table. During periods of heavy rain the ground often becomes soggy because the top layers of the soil become saturated. As the water settles in the soil, the zone of saturation may move upward in the soil profile. And in drought, the water table drops. This is why shallow wells go dry during periods without rain.

When water is pumped to the surface, the water table around the well lowers according to how much water has been withdrawn. The drop creates a cone of depression. In this way, nearby water in the aquifer is pulled into the area around the well.

In normal conditions, groundwater is continually replaced. But if water is pumped out faster than nature can replenish it, the cone of depression becomes very large and the water table drops. This situation can be likened to an overdrawn checking account.

Some underground water is a rather rare commodity called "fossil water." Here, the aquifer is completely encased in nonporous rock, and holds water that was trapped eons ago. Once the water runs out, there will be no more.

The quality of groundwater depends on a number of variables, but the greatest factor is the rock formation of the aquifer. If rain is acidic, it can dissolve certain kinds of rock so groundwater may contain more minerals and salts than some surface waters. As the rain percolates through the soil, it picks up sodium, calcium, sulfate, magnesium, carbonate, chloride, and nitrate and other easily solubilized minerals. When groundwater contains high concentrations of these substances, it is called mineralized.

In addition to the kind of rock within the aquifer, quality depends on the location of water within the aquifer, and where the aquifer stands in relation to other bodies of water. For example, a shallow well will tap into "young" water that only recently percolated into the ground. It reflects the quality of the rain it came from. Moreover, the quality of young water will vary with the seasons. Deeper wells pierce into a more stable supply of water. Because this water has had longer contact with the surrounding rocks, it may have a very high mineral content. Sometimes wells hit water that is in contact with ancient buried forests or swamps. This water is high in iron and sulfur, and may stain laundry.

But a more typical problem with wells is that, as the underground water is pumped out, the cone of depression can draw in water from surrounding areas. In this way, polluted water from a nearby creek or stream can be drawn into the well.

Half of the people in the United States drink water from underground sources. But because groundwater is concealed and confined, and not so dramatic as stinking lakes and rivers, and the sight of fish floating belly up, we tend to overlook it.

Once an aquifer is polluted, perhaps by leachate from a septic tank or a landfill, the only way to stop the contamination is to remove its source. Often, the owner of a well must enter into litigation to have the offending source removed.

The long-range effects of groundwater pollution are more serious than surface water pollution because it often takes many years to flush waste out of an aquifer. Most often, people who have bad wells simply choose to abandon them rather than try to cleanse them.

Causes of Groundwater Contamination

In most cases of groundwater contamination, the source is leachate from septic tanks or cesspools. Careless planning by sanitary engineers, excavators, or unknowing neighbors often causes the shutdown of a productive well when fluid from a waste disposal site flows into an aquifer. Each year about 800 billion gallons of domestic wastes are pumped into the ground. Fortunately, not all of it reaches the water table.

Sometimes the cause of a water quality problem is not human waste, but animal waste. Take, for example, the well used for coffee water in the town of Crosby, North Dakota. These good neighbors favored this well because the water had real body. Those who used it in their percolators claimed they made the best coffee in the state. And when health officials confiscated the handle of that well pump because the water was unsafe to drink, those residents became angry. But, it seems, the water had high concentrations of dissolved solids, sulfates, nitrates, and chlorides. The reason for this excess is that leachate from a livery stable entered the water. Traces of horse manure that had been buried thirty to sixty years earlier simply migrated to the well. Apparently this "secret ingredient" made for a mighty tasty cup of coffee![5]

Sometimes groundwater is contaminated by natural causes, such as an excess of minerals, or an influx of salt water. Mineralization results from the slow erosion of rock in the aquifer and can cause water to taste salty or bitter. Highly mineralized water is not usually suitable for drinking. In parts of Texas, Colorado, and the Dakotas, the groundwater has such a high concentration of fluoride that people have brown and spotted teeth. And in parts of Nebraska, water is so mineralized it looks like coffee when it is drawn from a well. Natural mineralization increases with the depth of a well, and eventually a depth is reached where the water is no longer potable.

A second cause of natural deterioration of groundwater is the intrusion of salt water. If water is removed from an aquifer faster than it is replaced, salt water is often drawn into the well. This situation is most common in coastal areas such as Long Island and Florida. If less water is drawn from the well, the salt water eventually will be flushed out by fresh. However, the process is extremely time-consuming and artificial recharging may be necessary. In this process, barrier wells are dug along the coastline, and fresh water is pumped into those wells to keep the salt water at bay.

But by far the greatest contamination to groundwater is the result of human activities. Human and industrial wastes are routinely buried in the ground, where they can slowly leach poisons into the underground water supply to cause serious health and economic problems.

Each year, more than 1,700 billion gallons of waste liquids are placed in the ground. About 800 billion gallons of that amount are from domestic septic systems. Other wastes are stored on the surface of the land, in landfills and dumps. These may soak into the soil, to be flushed by a heavy rainfall into the groundwater.

One of the most direct causes of water contamination is from injection wells, where wastes are pumped thousands of feet beneath the ground's surface. Leakage of wastes from the well casing into shallow ground is fairly common. Fresh water aquifers have been used for waste injection, but saltwater aquifers are more commonly chosen. Deep well injection of liquid wastes has become a common practice since 1960, particularly for the chemical, petrochemical, and pharmaceutical industries.

In some areas, liquid wastes are stored in lagoons or evaporation pits where they dry up. However, while the liquid is evaporating, it also is seeping into the soil. An added problem is that the walls of these lagoons sometimes collapse during a rain. In one case, the overflow from an evaporation pit killed all the surrounding plant life for more than 350 feet.

A recent report commissioned by the Environmental Protection Agency showed that one or more hazardous organic or inorganic substances had contaminated the groundwater at each of fifty different industrial waste disposal sites. Among the contaminants found in the groundwater were selenium, arsenic, chromium, lead, cyanide, copper, nickel, and barium. Organic contaminants included polychlorinated biphenyls, chlorinated phenols, benzene, and other solvents. More than half of the fifty sites showed concentrations of one or more of these pollutants in excess of federal drinking water standards—but all *fifty* are contaminated!

There are no federal standards on any of the above organic chemicals. Noteworthy is the fact that, based on recent EPA tests, certain groundwater supplies are considerably more contaminated than any surface water supply that has been monitored. For example, well water in certain parts of New Jersey has shown concentrations of benzene and chlorinated solvents in the several thousand parts per billion (ppb) range, whereas the same chemicals are present at concentrations less than 10 parts per billion in New Orleans, Cincinnati, Philadelphia, and Miami (cities which utilize surface water for drinking).

A more common and familiar pollution of groundwater results from the leaching of septic tanks and cesspools. This problem is most common in rural areas, where individual homeowners dispose of their domestic waste in the backyard. Usually cesspools and septic tanks give no treatment at all to protect the groundwater supply, yet these same homes have wells that tap underground aquifers for their drinking water.

In most towns, domestic wastes are treated by municipal sewerage plants. Plants may dispose of domestic waste in shallow wells or deeper injection wells. There have been fewer reports of contamination of groundwater from the latter.

The increasing use of fertilizers and pesticides in agriculture is reflected in the well water we drink. Chief among the contaminants are nitrates—a decomposition product of ammonia fertilizer. Throughout America's great "bread basket," high levels of nitrates in well water can be traced directly to the overuse and mismanagement of chemical fertilizer. Making matters worse, erosion and the resultant loss of organic matter in the topsoil, that would store this same nitrogen in an insoluble form, cause nitrogen to run off into surface waters.

This nitrate-rich water can lead to a serious disease in infants under four months old. Children suffering from it are commonly called "blue babies." There also may be a link between high nitrogen content in water and gastric cancer due to the formation of nitrosamines in the stomach.

Phosphates, chlorides, silicates, pesticides, and herbicides are also detected in the groundwater beneath crop land.

Another agricultural practice contributing to groundwater contamination is irrigation. In heavily irrigated areas such as the Central Valley of California, water not absorbed by crops picks up fertilizers and pesticides and carries them down into the groundwater. Generally, crops are overirrigated—with as much as 50 percent of the irrigation water returning to the ground. To make these problems worse, sometimes soluble fertilizers are added directly to an irrigation sprinkler system. But when the irrigation pump is turned off, the fertilizers easily are sucked back into the well. These irrigation practices are believed to be the worst cause of groundwater contamination in agricultural areas that depend upon irrigation.

There are various additional ways groundwater can be spoiled. These range from buried gasoline tanks that leak, to the paving of suburbia—pavement that keeps rainwater from replenishing aquifers that must meet the demands of burgeoning Levittowns. In the colder areas of the country, groundwater is often contaminated by salt that leaches into it after highways are deiced. The salt problem is not a small one. For example, in Maine alone, about 22 tons of salt is dumped on each highway mile each year!

Summary

Our water supply is constant. There never will be more water, there never will be new water—only the same microscopic molecules used or stored, evaporated or frozen, recirculated and reused by human and beast, from now until the world ends.

Water is a universal solvent that can dissolve and carry almost every substance it touches. And everything it touches will affect its quality to some measure. In recent years, water has been burdened with increasingly greater amounts of contaminants and pollutants. We can realize the *total* burden only when we stop thinking of oceans, rain, lakes, and the ice that tinkles in our glasses as separate entities. They are all one, changing from form to form without our notice. And without our care.

Water is the lifeblood of this planet. And whether it is found in the vastness of the oceans or in the very pores of the rocks beneath our soil, it is our obligation to care for it as the treasure it truly is.

NOTES

1. "Acid Rain: A Serious Regional Environmental Problem," *Science,* June 14, 1974, pp. 1176–1179.

2. Gene E. Likens, F. Herbert Bormann, and Noye M. Johnson, "Acid Rain," *Environment,* March, 1972, pp. 33–40.

3. H. M. Seip and A. Tollan, "Acid Precipitation and Other Possible Sources for Acidification of Rivers and Lakes," *The Science of the Total Environment,* Vol. 10, 1978, p. 259.

4. Eutrophication is likened to an aging process during which a lake becomes so rich in nutritive compounds that algae and other microscopic plant life become superabundant. Certain species replace others, reducing diversity. The lake is in a sense being "choked"—a process which will eventually cause it to dry up and "die."

5. Wayne A. Pettyjohn, "Good Coffee Water Needs Body," *Water Quality in a Stressed Environment,* ed. by Wayne A. Pettyjohn (Minneapolis, MN: Burgass Publishing Co., 1972), pp. 194–199.

4 What's in <u>Your</u> Water?

The average person has no idea of the true condition of his drinking water. But then, neither do the experts who process that water. The scientific, economic, and environmental communities have pigeon-holed the standards that must be met before water can qualify as "good." Each group looks out for its own particular interests. In that way, traditional water experts feel that water is "good" if it is free of bacteria and viruses. Environmentalists won't consider it good unless it is also free of potentially harmful chemicals. Economists, leaving health considerations to others, look at the cost-effectiveness of each water purification plant. As a result of this kind of independent evaluation, we live with a hodgepodge of judgments, standards, and policies. And, depending upon whom you ask, your drinking water may qualify as good or it may not.

In the hot seat are the water commissioners of our municipalities, who must answer to the scientists, ecologists, engineers, and economists, as well as to the public. At times, these commissioners cannot satisfy all factions. For example, the feisty water commissioner of Philadelphia, Carmen Guarino, performs a minor miracle every day. Overseeing a small feifdom of water treatment plants, he treats water from a source that may have been used as many as seventeen times before it gets to Philadelphia, changing it from sewage to drinkable water.[1] This city's water supply is drawn, in part, from the Schuylkill

River. Up-river towns also use this supply, and return what has been used as waste. When the water is drawn into the Philadelphia system, it has already been used by other cities, such as Reading.

Guarino bridles at the Environmental Protection Agency's recommendation that Philadelphia's water should receive added treatment by filtering it through beds of granular activated carbon so that the trihalomethanes—suspected carcinogens resulting from the chlorination process—will be removed. He says that while a very small number of people may get cancer from drinking Philadelphia public water (he estimates one death per year from these cancers) he feels the cost of carbon beds is not justifiable. He stated in *Civil Engineering* magazine, "A public official has to sacrifice that life; the money can be better spent elsewhere. The same money spent on improved street and highway safety or more effective crime prevention might save many more lives."[2]

In deciding whether to treat his drinking water himself, a homeowner has to make the same judgments a water commissioner makes. How will the water affect the health of his family? How much will it cost to clean up his water? Is the expense justifiable, considering the other demands on his budget? Hopefully, the individual will make a wiser decision than Philadelphia's water commissioner. People expect their food, air, and water to contribute to their health and well-being, not to undermine it. But we are not willing to invest in clean water.

Health is not something hyped by Madison Avenue, and homeowners are likely to decline a domestic water treatment system that costs from $300 to $500. We are accustomed to spending that amount on a washer and dryer for clothes, but not on treatment for clean water.

Testing the Water

To get a clear picture of the quality of your water, you'll have to test it to see if harmful elements are present. Begin the way any good water commissioner would, by taking a sample and having it tested for bacteria. On your own, you can evaluate its color, odor, taste, and other physical characteristics. One or more of these elements may indicate problems that require further testing. You will want to find the pH of your water to determine if it is corrosive to your plumbing (and your health). Water should also be tested for the presence of nitrates, for heavy metals, and perhaps even for the presence of organic chemicals.

Some of these tests can be done in your own home with kits purchased from chemical companies. Other tests require sophisticated equipment and/or procedures to give accurate results; if you decide you need these tests, you will have to send your sample to a laboratory. The cost of tests varies widely—from $1 to $50 or more—depending on the information required. Appendix D lists the names and addresses of the health departments in each state which can supply information on how you can have your own water supply tested.

Bacteria

Let's apply some educated guesswork about the bacterial quality of your water. A 1970 survey of 969 communities showed that 41 percent did not meet minimum federal drinking water standards. According to a United States Public Health Services Community Water Supply Study, the smaller the system, the greater the probability the water is substandard. Of those systems that served fewer than 500 people, half did not meet the minimum standards and nearly one-quarter exceeded the mandatory limits for contaminants.[3]

According to statistics, you're even worse off if you are one of the 50 million people in the United States who depend on their own well or cistern. Federal and state studies *conservatively* estimate that 40 percent of these systems may be contaminated. In other words, it's entirely possible that 20 million people are drinking polluted water. Other studies during a twenty-five-year period show that of the 360 known outbreaks of diseases or poisonings caused by water, over 70 percent involved private water supplies.[4] Those, of course, are the reported outbreaks. The EPA estimates the real number may be ten times that.

So, right off the bat, you can see the chances are good that your water is contaminated with bacteria.

Don't dismiss the possibility simply because you have not been ill. Some people drink contaminated water and never become sick from it. A family can build up an immunity to familiar bacteria, and never suffer any ill effects. It's common for bacteria to seep from a family's septic system into their well, recycling the bacteria they are already immune to.

So, the first step in evaluating your water—especially if you supply your own—is to have it tested for bacteria. These tests are usually done by the local or state health department without cost; if you are charged, the fee will be nominal.

Contamination is gauged by a count of coliform bacteria. These are bacteria found in the intestines of humans and other warm-blooded animals, and in the soil. While they are normally harmless by themselves, their presence in water may indicate the presence of human or animal waste. Scientists presume that a water supply showing the presence of coliform bacteria is unfit to drink. The source of the pollution should be found and corrected.

Tests for the presence of bacteria are done by taking a sample of the water to be tested and placing it on a medium suitable for bacterial growth. The sample is diluted sufficiently so that individual bacterial cells are separated from each other when they are placed on the growth medium. Each bacterial cell grows and multiplies to form a colony of cells which can be easily counted. Since each colony arises from one bacterial cell in the water sample, an estimate of the number of bacterial cells in the initial sample can be made. The standards for deciding whether a specific bacterial level (count) indicates contamination is somewhat arbitrary but is based upon experience which shows that drinking water with certain levels of bacteria leads to illness. In other words, it can make you sick.

Taking a Water Sample

Ask your local health department for a sterile sample bottle and a list of instructions. Most likely, the instructions are similar to these:

1. Use a sterile sample bottle provided by the laboratory. Be careful not to touch the cap or the mouth of the bottle with anything. Do not open the bottle until you are ready to fill it. Do not rinse it.

2. Inspect the outside of the cold water faucet to be sure there are no leaks around the handle. If there is leaking, take the sample from another faucet.

3. Remove any attachments, such as a hose or faucet aerator. Clean the outside of the faucet.

4. Let the water run at full force for about two minutes in order to clear the pipes and storage unit before you collect the sample.

5. Reduce the water flow to one-third of full force and run again for about two minutes.

6. Open the sterile bottle and fill it to within one-half inch of the top. Be sure the bottle is held so that water touching your hands won't enter the bottle. Put the cap on the bottle immediately.

7. Mail or deliver the sample to the laboratory as soon as possible. It is best to keep the sample in an ice cooler for storage during transport. A sample more than twenty-four hours old may not give accurate results.

In a week or less you should receive the results of the bacterial testing, telling whether your water is polluted or not. Some labs give the exact coliform count. There are two standard methods of determining the number of coliform bacteria in a sample: the multiple tube fermentation technique and the membrane filter technique. It doesn't really matter which method was used—both are accurate—but the allowable number of coliforms are different in each test.

The multiple tube test will report the coliform count in your water sample as the most probable number, indicated as MPN on the form sent to you. The count will not be exact, but will indicate whether the number has exceeded the EPA standards for drinking water. In this test, less than 2.2 coliforms per 100 milliliters of water (a milliliter is one one-thousandth of a liter, or about one-fifth of a teaspoon) is permissible. If three or more of the multiple tubes have MPNs above this level, the water is considered unsuitable for drinking.

The membrane filter technique tests positive if more than one coliform colony per 100 milliliters shows up.

Below are two standard bacteriological reports, one from South Carolina, the other from Missouri. The South Carolina form reports the coliform count per 100 milliliters using a membrane filter technique (MF). It requires you to state the source of supply (well, spring, or cistern), the type of well (dug, bored, drilled, or jetted), the well depth, and the exact location from which the sample was taken. The form used by the state of Missouri is one of the simplest available, just asking for the date and place of collection, name of collector, and source of sample.

But some states ask for a bit more information. For samples from a private well, they may ask whether the well is dug, drilled, or bored. They may also want to know the depth and circumference of the well.

SOUTH CAROLINA DEPARTMENT OF HEALTH AND ENVIRONMENTAL CONTROL ▪ OFFICE OF ENVIRONMENTAL QUALITY CONTROL

Date Collected	Time Collected	County		Rec'd By		Date Rec'd		Time Rec'd

Type of Supply	Individual Well	City-Town	Trlr. Pk.	Inst.	School	Bottled	Motel	Rest	Dairy	Pool	S/D	Other

Name of S/D or locality being sampled (omit if individual well)   Name of location of water plant (omit if individual well)

Type of Sample	Routine	Re-Check	Main Clearance	Well Survey	Fecal	Other

Lab	Columbia	Aiken	Charleston	Florence	Greenville	Lancaster	Remarks

Phone Number _____   Name and mailing address of person/firm to receive report

	(Individual Supply Only)
Name: _____	Source of Water
Street: _____	( ) Well ( ) Spring ( ) Other
City & State: _____	Type: ( ) Dug ( ) Bored ( ) Drilled ( ) Jetted
	Depth: _____

Sample Collected By: _____

EXACT LOCATION OF SAMPLING POINT MUST BE GIVEN   (Name and Address)

Lab Number	Sample No.	Location of Sampling Point	Cl Res'd	MF (Total) per 100 mls	MFC (Fecal) per 100 mls	Satis.	Unsat.	Verif'd.	NCG.	SPC 1 ml

WATER BACTERIOLOGY   Examined By _____ Date _____ Time _____
                                Reported By _____ Date _____

DHEC–1309 (Rev. 12/76)   Interpretations - Recommendations

Figure 4-1: Water Bacteriology Report Used in South Carolina.

Occasionally, a form asks what type of pump is used. If you do not have the answers to these questions, contact the builder of your home and ask him for the name of the company that installed your well. They will have records that supply all the needed information. If you live in a very old home, you may have to ask a well drilling company representative to examine your well to give you the needed information. In any case, the water sample form is easy to complete, and well worth the effort.

However, test results indicate the condition of the water only at the time of the sampling, and water quality can change rapidly. For example, if Barbara and Matt Coyle of Reading, Pennsylvania, had their water tested the day before the Schuylkill River flooded their well, the results may have declared the water safe for drinking. The day after the flood, the same water had 550 times the allowable number of coliform bacteria.

Therefore, it's only common sense to have your water tested periodically. Usually, once or twice a year is sufficient if the water tests clean each time.

M I S S O U R I   CHRISTOPHER S. BOND
                                   GOVERNOR

DEPARTMENT   OF
Natural   Resources   JAMES L. WILSON
                                           DIRECTOR
P. O. Box 1368   Division of Environmental Quality
Jefferson City, Missouri 65101   Phone: (314) 751-3749

REPORT OF BACTERIOLOGICAL EXAMINATION OF WATER SAMPLES
Private Water Supply

Samples collected by —

Date of collection —

Place —

Date reported —

Laboratory Number	Source of Sample	Number of Coliform Organisms Per 100 ml. portion

Based upon bacteriological standards, the above results indicate that, at the time the sample was collected, this water was _____ for drinking purposes.

The safety of a water supply depends upon proper construction and protection against contamination. A favorable bacteriological analysis alone should not be accepted as conclusive evidence of the safety of a water supply unless a survey of the supply indicates no sanitary defects. It is recommended that a water supply used for drinking purposes be analyzed routinely.

The above examinations were made by the membrane filter technique in accordance with the latest edition of Standard Methods for the Examination of Water and Wastewater.

Figure 4-2: Standard Bacteriological Report for Water Testing Used in Missouri.

If you want to try testing yourself, the Hach Chemical Company, P.O. Box 389, Loveland, Co 80537, offers a kit. Designed for schools and other educational institutions, it requires very little training to get accurate results. However, the initial investment—which includes an incubator—costs about $50. Unless you live in extreme isolation, it's probably better and cheaper to have the local health department do the testing for you.

In addition to testing for bacteria, many local health authorities routinely will test for hardness, pH, and, in agricultural states, for nitrates where suspected in drinking water.

Hard and Soft Water: The Inside Story

While most people in the United States contend with hard water in their homes, few understand what makes water hard or soft.

Hard water contains large amounts of dissolved minerals, especially calcium and magnesium, which it picks up while trickling through underground deposits of dolomitic limestone. Technically,

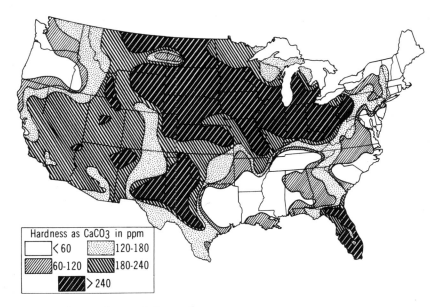

Figure 4-3: Hardness of Groundwater.
(Adapted from Water Atlas of the United States, *by James L. Geraghty et al. [Port Washington, NY: Water Information Center, 1973], plate 42, by permission of the publisher.)*

The map shows the general pattern of hardness of groundwaters in the United States. The largest hard-water areas are underlain by limestone. Only a few sections of the country, notably the New England states, some areas along the Atlantic and Gulf Coastal Plains, and large parts of the states of Oregon and Washington, fall into the soft groundwater category.

The patterns shown on the map do not necessarily mean that all groundwaters in a specific area have exactly the same hardness.

water is defined as hard if it contains more than 75 milligrams of mineral particles in a liter.[5] But water hardness is often measured in "grains" or grains of minerals per gallon. Water is considered *slightly hard* if it contains 1 to 3 grains per gallon; *moderately hard* if it contains 3 to 6 grains per gallon; *hard* if it contains 6 to 12 grains per gallon; *very hard* if it contains 12 to 30 grains per gallon; and *extremely hard* if it contains 30 or more grains per gallon.[6]

In other words, a boiled down gallon of moderately hard water would leave behind a mineral deposit equal to the size of an aspirin tablet (5 grains). That doesn't sound like much. But if you ever moved from an area where you enjoyed soft water to one where the water supply is hard, you know what a difference these dolomitic minerals can make.

The bathtub develops a ring that wears longer than most wedding rings. There's no satisfying layer of soap suds in the washer. Hair doesn't squeak after shampooing. Clothes made from synthetic fabric look dingy. The bottom half of a double boiler appears to have terminal psoriasis, the chalky innards of a kettle flake into your tea, and jars processed in a canner emerge looking scabrous and unhealthy.

If yours is a hard water area, you may have been visited by a representative from the local water conditioning company. Representatives offer to test hardness with a scientific-looking kit, probably for free. Accompanying the free test is a free talk on the evils of hard water. As hard water evaporates, the story goes, it leaves behind a gritty mineral residue on the interior of the hot water heater, water pipes, radiators, and even on the inside of your bathroom cup. Conditioning companies suggest that it's cheaper to install a home water softener than to replace the household plumbing.

Soft water truly is an attractive commodity. It eliminates soap curd and detergent deposit from clothing. You need less soap for laundry and cleaning. White mineral scales no longer deposit on glasses, dishes, and utensils. Soft water rinses away grime faster, and leaves no residue. It allows for luxurious bubble baths and leaves hair clean and shiny. It provides a longer life for household appliances that use water, and it protects expensive plumbing and heating equipment.

The trouble is, although soft water seems to be a plumbing panacea, it promises few benefits for personal health. Just by definition, soft water lacks calcium and magnesium—two minerals which are essential to bodily functions and optimum health. Of course, water is not our sole source of these minerals; we get much of our daily allowance in the food we eat. But that doesn't discount the glaring evidence that, for whatever reason, people who drink soft water suffer more heart attacks and strokes than people who live in hard water areas.

About twenty years ago, a lone researcher discovered that deaths due to stroke occur more frequently in areas of Japan where water is soft. Of course, at that time it was too early to even suggest that soft water might contain a dangerous substance, or that the hard water had some protective factor. Or, for that matter, whether the water had anything at all to do with the statistics.

Yet study after study uncovered similar correlations with regard to heart attacks. In England and Wales. In Canada. In the United States. In Norway. Always the same—the harder the water in a given area, the lower the cardiovascular death rate. And conversely, the softer the water is, the higher the toll of heart attacks.

Not only that, more recent studies have disclosed that the proportion of "sudden death" is significantly higher in the soft-water areas. Sudden death refers to a complication which may follow a heart attack when the heart suddenly goes into rapid uncontrollable spasms called fibrillation. Since these random contractions are not capable of pumping blood, death ensues.

In England, for example, death certificates were obtained for men and women aged thirty to fifty-nine years old in six towns with soft water and six towns with hard water. Not surprisingly, for every age group, death rates from heart disease were higher in the soft water areas. However, it was also noted that the proportion of sudden deaths was consistently higher in the soft-water towns than the hard-water areas when subjects were classified by age or social class.[7]

Still, it's hard to believe that the very minerals which are capable of clogging water pipes can prevent deaths due to clogged arteries. Or that something which sounds as innocuous as *soft* water might harbor some hidden health hazard. But, strange as it seems, both may be true.

For example, there's more than one reason to believe that the magnesium found in abundance in hard water supplies may somehow toughen the heart against sudden disturbances in heartbeat.

First of all, heart muscles of persons living in hard-water areas (where cardiovascular death rates are low) tend to contain more magnesium than those of persons living in soft-water locales. In heart muscle samples collected at autopsy from eighty-three cases of accidental death and analyzed for mineral content, Terence W. Anderson, M.D.,

Ph.D., of the University of Toronto's Department of Preventive Medicine and Biostatistics, reported that magnesium content varied significantly between hard- and soft-water areas. In fact, magnesium concentrations were 7 percent lower in the heart muscles of subjects who lived in the soft-water areas of Ontario.[8]

What's more, heart muscles of heart attack victims contained, on the average, 22 percent less magnesium than the muscles of healthy persons who died accidentally—a finding that echoes the results of an even more recent Israeli study.[9]

Magnesium has been shown over and over again to play an important role in the prevention of heart arrhythmia or fibrillation as it is involved in reactions which are essential to the contraction of heart muscle.[10]

But magnesium's role in heart disease doesn't stop there. A group of physicians at Georgetown University Medical Division and the District of Columbia General Hospital in Washington, DC, report that this important mineral has a decided effect on lowering blood pressure—another serious risk factor in heart disease.

Six patients with hypertension (high blood pressure caused by no known physical problem) and four persons with normal blood pressure were given magnesium sulfate intravenously for ten minutes. Afterward, their blood pressures were taken, as well as measurements of cardiac output (the volume of blood pumped per minute).

The researchers noted that, while the magnesium had no effect on those subjects with normal blood pressure, it reduced the blood pressure of the hypertensive individuals by 10.8 percent. In addition, cardiac output in all subjects increased after the magnesium infusion. This means that the heart was able to pump more blood per minute, making it a more efficient unit. Furthermore, in those people with hypertension, magnesium managed to decrease the resistance in their arterial walls, which is the cause of increased pressure.[11]

Furthermore, a magnesium deficiency can predispose a person to a calcium deficiency. And a calcium deficiency may also be a problem in heart disease—which is why hard water, with its ample supply of calcium, may provide double protection against heart disease. In fact, a separate English study that analyzed water from homes in sixty-one areas of England and Wales found calcium—not magnesium—to be the number one common denominator in the hard water–low cardiovascular death rate association.[12]

Earlier, a group of South African physicians reported that heart attack victims tend to have lower blood levels of calcium and pass less calcium in their urine than healthy persons with no signs of heart disease.[13]

Mounting evidence suggests that yet another mineral, which usually occurs in small "trace" amounts in hard water, may prove to be a big factor in the link between water hardness and heart disease. It's called silicon. And Klaus Schwarz, M.D., of the Department of Biological Chemistry at the UCLA School of Medicine, probably did the most to illuminate the connection between this trace mineral and heart disease.

In one study, he noted the results of a well-known survey of heart disease mortality in Finland. Between 1959 and 1974, men in an area of eastern Finland died of coronary heart disease at a rate twice as high as a group of men from western Finland. The usual risk factors, like smoking and obesity, couldn't explain the difference.

When a team of investigators led by Dr. Schwarz tested the water in the two areas, they made an important discovery: in the area where the risk of heart disease was high, the level of silicon in the water was extremely low. In western Finland, where the mortality rate was lower, the level of silicon was significantly higher.[14]

The possible importance of silicon is not too surprising when you consider that it is found in connective tissues such as cartilage, tendons, blood vessels, and similar structures throughout the body. Silicon, researchers think, may play an essential role in making these tissues strong and resilient.

For example, studies of human skin and aortas show that as connective tissues deteriorate with age, some silicon is lost.

More important, it has been found that the silicon in arterial walls decreases with the development of atherosclerosis.

If silicon disappears from artery walls as atherosclerosis progresses, it seems logical to wonder whether an insufficient intake of silicon isn't part of the problem—and whether the addition of this mineral to the diet may help prevent or treat the disease. A growing number of studies, in fact, point just that way.

Many studies, for example, have suggested that a high intake of certain high fiber foods may lower blood cholesterol levels and reduce the risk of heart disease.

When Dr. Schwarz analyzed the silicon content of many forms

of fiber, he found a revealing pattern. Sources of fiber with the demonstrated ability to lower cholesterol or prevent atherosclerosis—like alfalfa, rice hulls, pectin, and soybean meal—tested high in silicon. Cellulose, which has no protective effect, tested low. And in wheat bran, which produced contradictory test results, he found uneven amounts: three different readings from three different samples.

Obviously then, soft water's relative deficiency of essential minerals makes it a poor investment toward better health. If that were its only shortcoming, however, the solution would be simple. To make up for the minerals missing in your water supply, you could make an extra effort to concentrate on high fiber foods as well as foods crammed with calcium (like cottage cheese, yogurt, and skim milk) and magnesium (such as whole grains, spinach, beans, and nuts). Or supplement your diet with mineral tablets. In fact, if you pick up a bottle of dolomite, you'll be tapping the same source of calcium and magnesium that hard water supplies do.

Unfortunately, though, this is only the beginning of soft water's sorry story. For one thing, naturally soft water tends to be more acidic than hard water and acidic water is better able to corrode pipes, leaching hazardous metals such as cadmium, lead, and copper into home water supplies. (More on this in Chapter 5, "The Elusive Ingredients.") But in view of the correlation between soft water and increased rate of stroke and heart attack deaths, it's interesting to note that cadmium is linked to high blood pressure, a major risk factor in cardiovascular disease. A St. Louis study showed that patients with high blood pressure had fifty times as much cadmium in their urine as people with normal blood pressure.[15] And, in Kansas City, Kansas—where the cadmium content of drinking water is three times that of Kansas City, Missouri—there is a higher incidence of high blood pressure and many more deaths due to cardiovascular disease.[16]

On the other hand, the calcium in hard water has been shown to limit the internal absorption of lead, cadmium, and zinc.

"Softened" water (that is, hard water which is made soft through the removal of calcium and magnesium) is no more acidic than it was before the process. But, it has another drawback, salt.

Water softeners, such as the type installed in home plumbing systems, work by a method called "ion exchange." It's a complicated procedure. But what they do, quite simply, is swap electrically charged particles of salt (sodium and chloride ions) for the water's calcium and magnesium. Water is directed into a tank filled with sodium-charged plastic beads. When a magnesium or calcium ion contacts a bead, they are drawn together like magnets. A sodium particle is released into the water and the calcium or magnesium particle takes its place on the bead.

The amount of salt which eventually ends up in your water supply depends, of course, on the amount of calcium and magnesium it had to begin with. The more grains of hardness in the water, the more salt in the softened water.[17]

Table 4-1: Salt Added When Water Is Softened

Initial Water Hardness (gr/gal)	Sodium Added by Softening (mg/qt)
1	7.5
5	37.5
10	75
20	150
40	300

Source: James L. Gattis, *Water Conditioning* (Fayetteville, AR: The Cooperative Extension Service, 1973), p. 11.

Generally speaking, Americans consume about five to fifteen times the amount of salt the body needs to function. And because salt is linked with high blood pressure and fluid retention, the added burden placed on the body by drinking salty softened water may cause health problems. In fact, two University of Massachusetts researchers have found that sodium in drinking water appears to increase blood pressure rates in persons as young as high school age.[18] Beyond that, softened water also creates an added burden on the ground that eventually receives it.

For these reasons, water softeners should not be installed without careful consideration. Perhaps the most important question you need to ask is, "How hard is my water?"

You can test your own water for hardness. One of the most popular test kits among both water conditioning dealers and individuals is the kit for hardness measurement. It provides you with a quick accurate method of checking the efficiency of your water softener. Tests are relatively easy to perform, utilizing a simplified version of the titration process done in larger laboratories. The hardness indicator is first added to the sample, followed by the addition of the titrating solution, one drop at a time. When the solution turns colors, stop adding drops of the titrating solution. The number of drops you have used to bring about the color change is equal to the hardness in grains per gallon.

The largest manufacturer of water test kits is Hach Chemical Company, P.O. Box 389, Loveland, CO 80537, which has a water hardness test kit available for under $6.

If, after testing, you find that your water is only moderately hard, pass up the softening system. If, on the other hand, you discover that your water hardness is as you suspected—the source of all your plumbing bills—consider having a water softener hooked up to your hot water system only. After all, hot water pipes are most vulnerable to the scaling residue of minerals left in the wake of water evaporation. Besides, laundry, bathing, dishwashing, cleaning, and home heating can benefit from the removal of the minerals that cause hardness, while at the same time, the family can enjoy drinking and cooking without added sodium.

You don't need softened water (and the resultant maintenance and expense) for watering lawns and gardens, or for flushing the toilet —and these are the only other cold water outlets in your home. In the kitchen sink or clothes washer, a mixture of softened hot water and hard cold water will give you satisfactory results.

Nitrates

Nitrates are naturally occurring substances found in our vegetables, our water, and our soil. Their presence is potentially harmful to humans and especially to young babies.

The disease that affects infants is called methemoglobinemia, and the children who suffer with it are commonly called "blue babies." What happens is that the nitrates—which are relatively harmless—are changed inside the baby's stomach into *nitrites.* The alkaline nature of a baby's stomach, especially those younger than four months of age, allows the growth of a type of bacteria that brings about this change. Nitrites then change the hemoglobin in the baby's blood so that it no longer is able to carry oxygen to the body's tissues. The result is that the tissues suffocate from lack of oxygen, and the baby turns blue. The disease can be fatal.[19]

Because infants are so susceptible to nitrites, no drinking water containing more than 45 milligrams of nitrates per liter should be fed to them. Unfortunately, boiling the water just concentrates the level of nitrates. Public health departments routinely suggest that mothers switch to bottled water for infants if there is any evidence of a high nitrogen content in tap water.

A second health problem related to nitrates in drinking water is caused by the formation of nitrosamines, which can cause cancer. Scientists are uncertain about how likely this is to occur, and have made nitrate research a priority item in the coming years. They speculate that a series of reactions within the human body may change nitrates into a cancer-causing substance. Apparently any nitrate we eat or drink can be changed into a nitrite. This nitrite then reacts with other substances present in the stomach to become N-nitroso compounds. More than 100 N-nitroso compounds have been tested on animals, and 75 to 80 percent have been found to cause cancer.[20] So far there is no definite evidence that these compounds will cause cancer in man, although there is no known reason why we should be immune.

Some studies have shown a relationship between drinking water high in nitrates and gastric cancer. People living in the little town of Worksop, England, drink water with a very high level of nitrates—90 milligrams per liter, as opposed to the 45 milligrams per liter allowed in this country. There, the incidence of gastric cancer was found to be 25 percent greater than among others who lived in similar control towns. For residents seventy-five and over, the incidence of gastric cancer was 100 percent greater in Worksop.[21]

Additional studies of the relationship between nitrates in drinking water and the incidence of gastric cancer were done in southern Colombia. High concentrations of nitrates—up to 80 parts per million— were found in drinking water supplies. Residents using these supplies for the first ten years of their lives had a greater prevalence of gastric cancer. Researchers studied the urine of these people, finding too that it contained high levels of nitrates.[22]

But these findings are preliminary. What scientists do know is that the largest amounts of nitrates come from our own saliva and from such vegetables as celery, potatoes, lettuce, melons, cabbage, spinach, and root vegetables. About 90 percent of the nitrates we consume come from food, especially vegetables, about 9 percent from cured meats (20 percent for nitrites), and less than 1 percent from drinking water.[23]

In water, the major sources of nitrates are septic systems, animal feed lots, agricultural fertilizers, and manured fields, as well as industrial waste waters, sanitary landfills, and garbage dumps.[24] Removing nitrates from water has presented real technical difficulties for municipal treatment plants. Ion exchange is the only method that has been used with any success, but it has been used by only one plant, Garden City Park Water District in Nassau County, New York.[25] Unfortunately, the plant has operated only sporadically since its origin in 1974, and provides sketchy results. A homeowner can remove nitrates by distillation, but more common solutions are drilling a new well, or using bottled water.[26]

Nitrates generally show up in well water, especially from shallow dug wells. While these concentrations are sometimes natural, reflecting the nature of the rock formations in the given area, usually they are caused by human or animal waste that has contaminated either surface water or aquifers.[27] Often there is a direct correlation between the depth of a well and/or its nearness to a septic tank and the amount of nitrates found in the well water.[28] The deeper the well, and the greater its distance from a septic tank, the less is the nitrate pollution. State and local health departments often test for nitrates without charge.

The pH of Water

Despite the advertisements for shampoo with an "adjusted pH," lots of people don't have the faintest idea what pH really means.

Actually, the symbol pH represents a measure of whether or not the water is acidic or basic (alkaline). When the pH is less than 7, the water is said to be acidic; above 7 the water is basic (alkaline), and if the pH is exactly 7, it's said to be neutral. Most groundwaters in the United States have pH values ranging from about 5.5 to 8. The most desirable range for water in the home is from 8 to 8.5.

Water with a low pH (acidic water) is more likely to corrode plumbing, causing leaks and other damage. Lead, copper, and other metals that enter the drinking water through this corrosion can have an adverse effect on health. (Other factors in water that can cause corrosion are oxygen concentration, electrical conductivity, and temperature.) Corrosion is a natural process: all metals have a tendency to

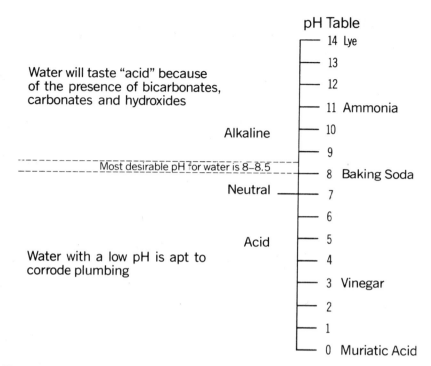

Figure 4-4: The pH of Water and Some Common Household Products.

revert to their original ore. Some metals, such as gold, are highly resistant to corrosion, but these are usually too expensive to be used in plumbing.

Groundwater that is hard is generally not acid, since the calcium and magnesium that make it hard are alkaline. But some groundwater —usually water that has not been drawn from a limestone aquifer, or water that has been influenced by acid rain—can be acid.

Recent data seem to indicate that corrosiveness may be the factor in water that accounts for increased cardiovascular problems in soft-water areas. (See Hard and Soft Water: The Inside Story, earlier in this chapter.) Cadmium leached from old galvanized pipes, lead from old lead pipes, and copper leached either from pipes or fittings may also have adverse health effects.

On the other hand, alkalinity is due to the presence of bicarbonates, carbonates, and hydroxides in water. Sometimes these substances can be detected by their acrid "soda" taste. Alkalinity can also be identified with common litmus paper, which changes from red to blue when alkalines are present.

It's important to know if your water is acid or alkaline, because it has a bearing on how the water can be treated. For example, if you have iron in your water, creating stains on laundry and bathroom fixtures, that iron cannot be removed from your water if it is acid. To buy and install a water treatment unit to remove iron, without first neutralizing the acid in the water, is simply a waste of money.

You can test the pH of your water easily. Kits can be purchased from the Hach Chemical Company, P.O. Box 389, Loveland, CO 80537, starting at about $7.50, and from fish tank or swimming pool suppliers. Or you can ask your local health department to test the pH at the same time it tests for bacteria.

Fireproof Water: It's No Joke

Most water becomes cloudy because silt or mud gets into it. Generally, these suspended solids are disturbing—they make kids say "Yuk!"—but they are not deadly. However, there is one deadly particle showing up in water tests with increasingly frequency. It is asbestos.

Inhaled particles have been proven to cause three serious diseases—asbestosis, a disease in which the lungs become clogged with fibers, making breathing extremely difficult; lung cancer; and mesothelioma, a cancer of the lining of the stomach or lungs that usually kills a few months after it is diagnosed.

But it is possible that asbestos causes additional health problems. Studies at the Mayo Clinic in Rochester, Minnesota, by Arnold L. Brown, M.D., and others have shown that asbestos fibers which have been inhaled or swallowed can travel once they get inside the body. This phenomenon may explain why asbestos workers experience higher rates of gastrointestinal cancer than the general public.

While asbestos naturally occurs in certain water supplies, it is turning up in more and more samples. Perhaps the most frequent single cause of this occurrence is the use of asbestos cement pipe to supply drinking water. Today, some 200,000 miles of this pipe are in use across our country.

Asbestos cement pipes came into common use after World War II. They were cheaper than the traditional cast iron pipes. And, because they are comparatively light, they were easier to transport and install. They became widely used during the boom years for real estate development, when tract housing sprouted up where corn fields and cow pastures had been. But now, according to the Environmental Protection Agency, those pipes are beginning to disintegrate. As they erode, asbestos fibers are released into drinking water supplies.

In areas where the water is acid (with a pH below 7), corrosion occurs rapidly. The EPA began analyzing water samples for asbestos back in 1976. These studies found massive quantities of asbestos in some systems. For example, in Cincinnati, the range of fibers went from 14 to 1,900 per liter; in Seattle, the number of fibers ranged from 400,000 to 1.5 million per liter; and in Pensacola, Florida, fibers ranged from 700,000 to 32 million per liter.

After these studies, the EPA concluded that "erosion of asbestos fibers from the walls of asbestos cement pipe may endanger community supplies."

Under the Clean Water Act, the EPA can regulate the asbestos levels in drinking water. However, the agency is hesitant to require towns and cities to rip up their pipes. Additionally, the EPA does not have positive proof that swallowed asbestos causes cancer—simply because most studies have concentrated on the effects of inhaled asbestos fibers. But a recent EPA study has shown that, where high levels of asbestos exist in the drinking water of certain parts of California, the cancer rate is above average, especially for cancer of the stomach, kidneys, and abdomen.

Some cities have acted to reduce the risk by slowing the disintegration of the asbestos cement pipes. This is done by making the water that passes through the pipes less acid, and thus less corrosive. By adding sodium hydroxide or lime, the pH of the water is raised. It is also possible to filter the asbestos out of the water, as was done in Duluth, Minnesota, during the period when that city's drinking water was heavily polluted by asbestos-laden mining wastes.

Eyeballing the Quality

In addition to municipal tests for bacteria, hardness, nitrates, and pH, you can do a little testing on your own. Pour a glass of water and take a close look and a good sniff. Sometimes just odor and appearance can tell you that your water needs more treatment than it's getting.

Does your water *foam* as it splashes into a glass? Foam can be an indication that detergent residue is getting into your drinking water. If your water source is a surface body—lake, stream, etc.—which is also used for waste water disposal, it probably isn't getting sufficient treatment to remove detergent, and maybe other matter as well.

If you have foamy water and provide your own supply, your septic system should be examined for leaking. Is its location far enough from your water source? Is your water source uphill from your septic system? It should be. Check your well to make sure it is properly sealed. If you have neighbors close by, their septic tanks should be equally suspect.

Is your water murky looking? *Turbidity* is an indication that clay, silt, metals, synthetic or natural chemical compounds, plankton, and microorganisms may be present in the water. It could also indicate the presence of sewage, industrial waste, and asbestos. These contaminants float in water and make it look cloudy. In groundwater, turbidity is usually caused by particles of soil that have percolated into the water table. Some well water becomes cloudy after a heavy rain because of this percolation.

Turbidity may mean the water filtration plant is not functioning properly. These plants usually get suspended matter out of water by coagulation, sedimentation, or filtration. However, sediment can get into the system even if the plant is operating at peak efficiency. Sometimes water leaves the plant in good condition only to become muddy inside cracked or leaking pipes. Water can also pick up color from rusty pipes inside your house or in your water heater.

Turbid water is unpleasant to look at, and even more unpleasant to drink. While a little silt will probably do you no harm, it can provide food and lodging for bacteria or pollutants such as pesticides, which are known to cause health problems.

The EPA also notes that turbidity can interfere with disinfection at treatment plants. In tests performed on turbid water containing iron rust and plankton, coliform bacteria were found even though the water had been treated with chlorine. The EPA has warned water treatment managers that, for chlorine to be totally effective in destroying germs, the turbidity in water must be reduced to nearly zero by means of coagulation and filtration.[29]

Does your water have a *color?* Truly pure water does not. Color can be caused by decaying organic matter—things like leaves and plants, primarily. If your water comes from a surface body, chances are the color has been supplied by Mother Nature's humus. While color does not make the water undrinkable, it is an indicator that the water has not been properly treated. Sedimentation and filtration should effectively remove all color from drinking water. Again, if you supply your own water, do some detective work to find out how decaying plant matter is entering your source.

Does your water have a peculiar *taste* or *odor?* The cause of off-taste and smells are many—ranging from industrial solvents to a dead animal decomposing in the cistern. Generally, the most common sources of bad taste and odor are chemicals that can add a funny flavor even if present in only a few parts per billion. These include formaldehyde, picolines, phenolics, xylenes, refinery hydrocarbons, petrochemical waste, phenyl ether, and chlorinated phenolics.[30]

Chlorine can add an unpleasant taste and smell to water. Removal techniques are discussed in Chapter 7.

Another common source of bad taste and odor are algae, which can make your kitchen or bath smell like a stagnant pond. Although algae are not usually associated with waterborne diseases, they can make drinking water pretty unpleasant. Fortunately, they can be removed by filtration and disinfection.

If your water smells like rotten eggs, there is hydrogen sulfide in it. Even in small amounts, this gas will make your water smell awful. And if you can smell it, enough is present to make the water unpleasant to use and to taste. Even small amounts in water are nauseating. Hydrogen sulfide is more often found in well water than in surface waters. Certain strains of bacteria feed on the sulfate and produce hydrogen sulfide as a metabolic by-product. If iron and manganese are present, the bacteria also will create iron sulfide and manganese sulfide. You don't have to be a scientist to know if your water contains iron or manganese sulfides—you'll know from the black slime on the interior of your pipes and water tanks. When this slimy build-up sloughs off, it can block pipes and fill your water with ugly black specks.

Because hydrogen sulfide is very acidic, it is also corrosive. In fact, concentrations of 0.5 milligram per liter can corrode iron pipes, filters, water tanks, and even concrete.

Unfortunately, well water usually is not tested for this gas when a well is installed. Consequently, many pumps and other submerged equipment made of ordinary ferrous metals quickly corrode away. If this has happened in your home, be sure to replace the equipment with corrosion resistant plumbing, like galvanized pipe. It is also possible to kill off the sulfur-eating bacteria and oxidize the sulfide ion into a less troublesome form. (See Chapter 7.)

Because of the dangers of hydrogen sulfide, and the taste and odor problems it causes, the standards for it are set at 0.05 milligram per liter. This standard is a recommended maximum contaminant level, presented as a guide for states and water suppliers.

Minerals also impart flavor to water, but not odor. Water high in minerals can taste metallic, but their presence generally is considered a health plus. Metals are discussed in detail in Chapter 5.

Of course, taste and odor are difficult to judge because of individual preferences. Some people just love the taste of their local water, while visitors may find it awful. Generally, we become accustomed to the water we drink and cease tasting it altogether. But if water should taste bad suddenly, it is a sure sign it needs testing.

NOTES

1. J. A. Borchardt and Graham Walton, "Water Quality," *Water Quality and Treatment: A Handbook of Public Water Supplies,* American Water Works Association, Inc. (New York: McGraw-Hill Book Co., 1971), p. 14.

2. *Civil Engineering,* September, 1977, p. 91.

3. Stanley S. Zimmerman, "An Approach to Provision of Rural Water Supplies," *The State of America's Drinking Water* (Washington, DC: North Carolina Research Triangle Universities and the U.S. Environmental Protection Agency, 1974), pp. 178–189.

4. Wilbur J. Whitsell, Gary D. Hutchinson, and Donald H. Taylor, "Problems of Individual Supplies," ibid., pp. 143–155.

5. *Drinking Water and Health* (Washington, DC: National Academy of Sciences, 1977), p. 439.

6. *Water Conditioning* (Fayetteville, AR: Cooperative Extension Service, 1973), p. 11.

7. Margaret D. Crawford et al., "An Epidemiological Study of Sudden Death in Hard and Soft Water Areas," *Journal of Chronic Diseases,* vol. 30, 1977, pp. 69–80.

8. Terence W. Anderson et al., "Ischemic Heart Disease, Water Hardness and Myocardial Magnesium," *Canadian Medical Association Journal,* August 9, 1975, pp. 199–203.

9. Abraham S. Abraham et al., "Serum Magnesium Levels in Patients with Acute Myocardial Infarction," *New England Journal of Medicine,* April 14, 1977, pp. 862–863.

10. Barbara Chipperfield and J. R. Chipperfield, "Magnesium and the Heart," *American Heart Journal,* June, 1977, pp. 679–682.

11. William J. Mroczek, Won Ro Lee, and Michael E. Davidov, "Effect of Magnesium Sulfate on Cardiovascular Hemodynamics," *Angiology,* October, 1977, pp. 720–724.

12. P. C. Elwood, A. S. St. Leger, and M. Morton, "Mortality and the Concentration of Elements in Tap Water in County Boroughs in England and Wales," *British Journal of Preventive and Social Medicine*, September, 1977, pp. 178–182.

13. "Calcium and Myocardial Infarction," *South African Medical Journal*, March 16, 1974, pp. 523–527.

14. Klaus Schwarz, "Silicon, Fibre, and Atherosclerosis," *Lancet*, February 26, 1977, pp. 454–457.

15. "Worldwide Controversy Persists on Link of Soft Water to Cardiovascular Deaths," *Internal Medicine and Diagnosis News*, vol. 5, no. 17, pp. 1–2.

16. "Water and Heart Disease: The Harder the Deadlier?" *Medical World News*, October 11, 1974, pp. 45–46.

17. *Sources of Sodium in the Diet* (Lombard, IL: Water Quality Association, February, 1978), p. 31.

18. *University of Massachusetts News Release*, June 27, 1978.

19. *Drinking Water and Health*, pp. 416–423.

20. Ibid., p. 421.

21. Ibid., p. 422.

22. Pelayo Correa et al., "Model for Gastric Cancer Epidemiology," *Lancet*, July 12, 1975, pp. 58–60.

23. *Drinking Water and Health*, pp. 414–415.

24. Ibid., pp. 411–412.

25. M. T. Gillies, ed., *Drinking Water Detoxification* (Park Ridge, NJ: Noyes Data Corporation, 1978), pp. 292–293.

26. George H. Klumb, "Nature of Water" (Conference Proceedings of the Farmstead Water Quality Improvement Seminar, sponsored by the American Society of Agricultural Engineers, Columbus, OH, October 5–7, 1966), pp. 5–8.

27. Ulric P. Gibson and Rexford D. Singer, *Water Well Manual: A Practical Guide for Locating and Constructing Wells for Individual and Small Community Water Supplies* (Berkeley: Premier Press, 1971), p. 25.

28. Ted L. Willrich and George E. Smith, eds., *Agricultural Practices and Water Quality* (Ames: The Iowa State University Press, 1970), p. 107.

29. Environmental Protection Agency, Office of Water Supply, *National Interim Primary Drinking Water Regulations* (Washington, DC: U.S. Government Printing Office), pp. 43–44.

30. A. A. Rosen and R. L. Booth, "Taste and Odor Control," *Water Quality and Treatment.*

5 The Elusive Ingredients

Once your water has been evaluated for bacteria and other disease-causing organisms, for turbidity, hardness, nitrates, and pH, you have completed the normal route for testing. But you still don't know if your water contains other pollutants that may be harmful to your health—pollutants such as heavy metals, pesticides, and even radioactivity.

Tests for these contaminants are not done as part of routine procedure. To detect minute quantities of organic chemicals, for example, technology is so refined and equipment so expensive that the test for each chemical may cost a laboratory more than $100 to perform. If you could find such a lab and have your water tested for the presence of all possible organic contaminants, the cost might be as much as $1,700.

Yet we know that tap water can contain substances like organic chemicals, or lead, copper, mercury, and arsenic. In fact, it is not at all rare to find these elements in household water, although usually in small amounts.

How can an individual know if these substances are in his own water supply? First, don't ask a laboratory to analyze your water and name everything that's in it. That isn't economically feasible to do for a water supply serving only one family. Still, you may be able to make an educated guess about what toxic elements could be in your water.

Dingy dungarees and plumbing problems are two sure signs that your water is studded with calcium and magnesium, the two most critical minerals in water. But there are many more minor minerals in both hard and soft water. And don't let their name fool you. Trace minerals—so-called because they account for such a small portion of the human body (less than 0.01 of 1 percent)—play an important role in the environment and can carry tremendous consequences for our health.[1]

Unfortunately, *trace* minerals are often misunderstood. When heavily concentrated in water supplies, for instance, they can be categorically chalked off as contaminants. This is understandable since practically all are harmful in large amounts and a few can be poisonous even at low levels. But the term "contaminant" is misleading since the same minerals which are harmful in high doses are actually necessary for bodily functions and can be beneficial to health in smaller doses.[2]

On the positive side, researchers like the late Henry Schroeder, M.D., one of America's most distinguished scientists, say that certain minerals are even more important to health than vitamins since they are absolutely essential to life and since they cannot be manufactured in living things.[3] But, by the same token, scientists warn that other hazardous minerals pose an environmental threat much greater than pesticides and herbicides because they cannot be broken down or decomposed by nature.[4] In a very serious way, you could say that trace minerals are a bit like a familiar nursery rhyme character. When they're good, they're very very good; but when they are bad, they are horrid.

For simplicity, then, we'll divide the trace minerals into two types —those which are harmful at all levels and those which are healthful in trace amounts. Lead, cadmium, mercury, and to a lesser degree arsenic belong to the hazardous group. Iron, zinc, copper, manganese, chromium, cobalt, molybdenum, iodine, and selenium belong to the second group, termed essential trace minerals. A few, such as barium and silver do not fit either category since there is little evidence that they are either extremely harmful or especially healthful.[5]

In any case, the key to optimum health would seem to be a careful balance of the essential trace minerals with a minimum exposure to the hazardous ones. Trouble is, it's often difficult to pick and choose the good from the bad. Trace minerals are everywhere. They're in the air we breathe, the food we eat, and, of course, in the water we drink.

We say, "Of course," but not too long ago water was considered an insignificant source. Food was thought to be the only worthwhile supplier of essential trace minerals to man.[6] Today, food is still number one but water is gaining greater recognition for several reasons.

For one thing, much modern food is no longer what it should be. It's been processed, refined, overheated, frozen, or otherwise stripped of essential minerals. Take flour for an example. White flour contains only 13 percent of the chromium, 9 percent of the manganese, 19 percent of the iron, 30 percent of the cobalt, 10 to 30 percent of the copper, 17 percent of the zinc, 50 percent of the molybdenum, and 17 percent of the magnesium found in whole wheat.[7]

What's more, it seems that the minerals in a bite of food are less readily absorbed by the body than those in a gulp of water. It's easy to understand why. In food, certain minerals are tightly bound together with amino acids and other substances and are not immediately available for nutritional purposes.[8] In water, most minerals freely float about in solution and can be absorbed through the stomach and intestinal walls with relative ease.

When you consider that the average person takes in about 2 quarts of water every day—and that the average home water supply can provide more than 10 percent of a person's requirements for major minerals like calcium and magnesium as well as trace minerals such as copper, iron, and zinc—it's clear that water can and does supply an appreciable proportion of the body's daily needs for a number of essential minerals.[9] Granted, food remains our primary supplier. But for those whose diet might be marginally mineral deficient, water's contribution could tip the scales from ho hum to optimum health.

As an added bonus, the minerals in your home water supply may also affect the amount of minerals available in the food your family eats. Food cooked in softened water (which is high in the mineral, sodium) will lose essential minerals which are leached out into the water. On the other hand, food cooked in hard water (without added salt) may actually pick some up.[10]

Problems arise, however, when water becomes tainted with hazardous minerals or inundated with essential ones—which, as we mentioned, can become dangerous in high concentrations. How does this happen?

Well, normally, minerals come from the soil, from decaying vegetation, and from rock formations below the ground surface.[11] When it rains, the rainwater (which tends to be soft and therefore slightly acidic) percolates through the soil, dissolving minerals in its path and carrying them deep into the water table. Likewise, surface runoff from rainfall may transport trace minerals to streams and lakes.

Of course, the types of minerals and their concentrations vary from place to place. For example, in areas of heavy rainfall or serious erosion, water concentrations of trace minerals tend to be higher.

Mining and industrial manufacturing also affect the type and concentration of trace minerals in a given location. If you live near a steel plant, iron foundry, or zinc mine, for example, your water supply may be burdened with a higher-than-average concentration of copper, zinc, or aluminum, not to mention a deadly dose of cadmium, mercury, or lead.[12]

Another possible source of trace mineral contamination in water is municipal solid waste landfills. In one study, the Environmental Protection Agency (EPA) examined five municipal waste landfill sites to see how they affected drinking water in the ground beneath them. These specific sites contained no industrial waste or sewage sludge, but

Figure 5-1: Movement of Leachate through the Land Phase of the Hydrological Cycle.

(Adapted from Water Quality in a Stressed Environment, *ed. Wayne A. Pettyjohn [Minneapolis, MN: Burgess Publishing Company, 1972], p. 138, by permission of the author and publisher.)*

Leachate from a landfill contaminates groundwater aquifer with dangerous trace minerals.

were dumping grounds for food waste, yard waste, glass, metals, plastics, rubber, liquid wastes, pesticide containers, paint cans, batteries, a wide variety of cleaning agents, dead animals, disposable diapers, grease, and hospital wastes.

Because 120 million tons of this kind of trash are disposed of in landfills in the United States each year, the potential for damage to underground water was evaluated. The study revealed that, in areas of the country where rainfall is limited, the problem is not very severe. However, where rain falls in quantities that exceed evaporation—where enough rainwater is left over after evaporation so that some seeps into the ground—the potential for pollution is great. Unfortunately, the areas tested with the largest number of municipal solid waste landfills coincidently had the greatest amount of precipitation.

Leachate was examined from landfills in Washington, Pennsylvania, Indiana, Tennessee, and California. What the researchers found was that leachate from the Pennsylvania site contained iron in quantities as high as 5,000 times the EPA limit. Leachate from the Indiana site contained nearly four times the allowable amount of cadmium. In Tennessee, leachate had as much as six times the allowable level of lead. These cocktails of contamination were also studied for copper, chromium, zinc, mercury, and selenium.

The water from these landfills creates health problems for the public because it can (and often does) seep into our underground water supplies. The damage it creates can be very dramatic.[13] For example, several residential wells in Aurora, Illinois, were contaminated by a landfill with a strong, black, foul-smelling runoff. People who depended on those wells were forced to use bottled water for sixteen months until a new supply was found. In the meantime, however, the contaminated water damaged sinks, faucets, and other plumbing fixtures.

Water treatment plants do remove some of these trace minerals from the water before it reaches you. But, interestingly enough, they also add some. All in the name of "purification," alum (aluminum sulfate) is added to coagulate with free-floating particles and cause them to sink, copper is added to control the growth of algae, and silver is sometimes added to filters to disinfect. Granted, we wouldn't want to drink water clouded by mysterious particles, or turned green by algae. But, neither do we want to drink water containing excessive amounts of such minerals as aluminum, copper, or silver.[14,15]

Studies have been done to compare the mineral content of untreated with treated water, and comparing treated water as it leaves the plant with treated water as it comes out of the tap. The studies are fascinating, because they show that mineral concentrations are often higher *after* the water has been treated than before. The treatment process was found to increase amounts of zinc, cadmium, iron, lead, molybdenum, nickel, cobalt, vanadium, copper, and aluminum.[16]

Table 5-1: Comparison of Mean Trace Mineral Concentrations in Untreated and Treated Waters in the United States (measured in micrograms per liter)

	Untreated Water[a]	Treated Water[b]
Aluminum	74	179.1
Barium	43	28.6
Berylluim	0.19	0.1
Cadmium	9.5	12
Chromium	9.7	7.5
Cobalt	17	26
Copper	15	43
Iron	52	68.9
Lead	23	33.9
Manganese	58	25.5
Molybdenum	68	85.9
Nickel	19	34.2
Silver	2.6	2.2
Vanadium	40	46.1
Zinc	64	79.2

Source: Data is taken from *Drinking Water and Health* (Washington, DC: National Academy of Sciences, 1977), pp. 210–211.

a. *1,577 raw surface waters in United States.*

b. *380 finished waters. These are not necessarily the same waters.*

Even worse, as water travels from treatment plant to your faucet tap, the problem of mineral contamination becomes even greater. The reason for this is that to a certain degree all water, whether hard or soft, can corrode copper, lead, or galvanized pipes. As the water eats its way into water mains and household plumbing, copper, zinc, lead, and cadmium leaches into your water supply.

A community water supply survey in 1970 evaluated the condition of water between plant and home (see Table 5-2). What they found was that 30 percent of the samples taken from the tap had mineral

Table 5-2: Community Water Supply Study of 2,595 Distribution Samples from 969 Public Water Supply Systems

	Standards		
	Limit (mg/l)	*Maximum Concentration (mg/l)*	*Percent Exceeding*
Cadmium	0.01	3.94	0.2
Chromium	0.05	0.08	0.2
Copper	1.0	8.35	1.6
Iron	0.3	26.0	8.6
Lead	0.05	0.64	1.4
Manganese	0.05	1.32	8.1
Silver	0.05	0.026	0.0
Zinc	5.0	13.0	0.3

Water samples below are not distribution samples, but finished water samples.

Arsenic	0.05	0.10	0.2
Barium	1.0	1.55	<0.1[a]
Selenium	0.01	0.07	0.4

Source: Leland J. McCabe et al., "Survey of Community Water Supply Systems," *Journal of the American Water Works Association,* November, 1970, pp. 678–687.

a. *This constituent was evaluated only on selected samples. The remainder were assumed not to exceed the limit.*

concentrations exceeding the government limits in existence at that time. In one case, for example, the cadmium content was eleven times greater than allowed in drinking water. In other samples, iron content was ninety times higher than acceptable; manganese, twenty-six times higher; and lead, thirteen times greater than set limits.[17]

Those with individual water supplies such as wells and cisterns are not exempt from this problem, because copper and lead household plumbing also can be a major contributor of metal pollution.

If you live in a hard water area (and do not have a water softener sharing space with your water heater) you may be somewhat exempt from this extra mineral dose. Calcium and magnesium tend to accumulate inside water pipes and create a natural lining which protects plumbing from excessive corrosion. On the other hand, if your water is naturally soft, you may have two strikes against you. One, your plumbing lacks any protective shield. Secondly, naturally soft water tends to be somewhat acidic, and acidic water is more corrosive. In Boston, for example—where the water is both soft and acidic—half the water samples taken from taps on Beacon Hill exceeded the lead limit.

All you need is a litmus paper kit (available in any pharmacy) to check the acidity or pH of your water. Take a look at the pH value of your water and check it against the figures shown in Table 5-3. As you can easily see, the higher the pH of water (or the less acidic), the less metal seems to be present in it. Cadmium, lead, and zinc corrosion seem to be slightly higher at the pH range of 7.0 to 7.4, but the corrosion of other metals is greater at lower pH measurements. Copper especially increases when the water measures at or below a pH of 6.9. If you have copper plumbing and your water's pH measures low, you can be fairly certain you have traces of copper in your water.[18]

For safety then, let your tap water run for about two minutes or so in the morning before drinking it. Because water has been in contact with the pipes overnight, there's likely to be a higher concentration of trace minerals from your plumbing system.[19] But by flushing your water lines, you can rid the system of the night's mineral-laden water.

Incidentally, the best time to take a water sample for testing is also in the morning just because mineral concentrations are highest then. After all, if your water is leaching dangerous minerals like cadmium and lead or excessive amounts of copper from your pipes, you ought to know about it. Besides, water is presumably checked for purity before it leaves the municipal treatment plant, but usually not when it comes out of your tap.[20] And *this* is the water you drink. If your water comes from your own source, for example, a spring or well, the importance of testing its quality is even greater since it probably hasn't been checked for metals at all before you drink it.

Of course, if you don't want to have your water tested for trace metals, you can employ some deductive reasoning and make an educated guess about which ones may exist in your personal water supply. For example, if you know that the water in your area happens to be soft and the pH tests below 8.0, you can suspect that the water is eroding your service lines and household plumbing. If your plumbing is copper, chances are you've got an abundance of that mineral in your water supply. If your plumbing is lead, you may have a more weighty problem on your hands.

You can also check your bathroom sink for signs of mineral contamination. A greenish stain just below the faucet is a sign of excessive copper. A reddish brown stain is an indication of iron.

Unfortunately, it's impossible to know the exact extent of trace minerals and mineral contamination in your home water supply without a formal test. And when you consider what a profound effect trace minerals can have on your health, it's silly to second-guess them. Rather, it would be wise to have your water tested. You needn't test for all minerals. But you should test for the hazardous ones like lead and cadmium, and the essential ones like copper and zinc which could be leaching into your water supply in unhealthy amounts.

State and local public health offices will often perform water testing—but usually for only a select number of minerals. For a more extensive trace mineral workup check the Appendix in the back of this book for a laboratory near you.

You can also have your water tested by the Soil and Health Foundation, a nonprofit organization which is concerned with environmental issues. For $25, they will test your water for the following: arsenic, barium, cadmium, chromium, cobalt, copper, iron, lead, manganese, mercury, selenium, silver, and zinc. In addition, they'll test for calcium and magnesium as well as pH. For a sample bottle and instructions, write to the Soil and Health Foundation, 33 East Minor Street, Emmaus, PA 18049.[21] Incidentally, to emphasize the importance of taking charge of your own water supply, the Soil and Health Founda-

Table 5-3: Mineral Levels Found in Distribution Samples from Community Water Supply Study

	pH to 6.9		pH 7.0–7.4		pH 7.5–7.9		pH 8.0 or more	
	Percent Exceeding Standard	Average (mg/l)	Percent Exceeding Standard	Average (mg/l)	Percent Exceeding Standard	Average (mg/l)	Percent Exceeding Standard	Average (mg/l)
Number of Samples	(425)		(556)		(550)		(595)	
Cadmium	0.0	0.001	0.5	0.008	0.2	0.001	0.0	0.000
Chromium	0.0	0.000	0.7	0.002	0.0	0.001	0.0	0.003
Copper	5.4	0.295	1.1	0.119	0.2	0.067	0.5	0.050
Iron	10.8	0.184	11.7	0.331	7.5	0.116	3.9	0.081
Lead	1.6	0.013	2.3	0.016	0.5	0.012	0.3	0.009
Manganese	9.9	0.026	11.7	0.033	6.8	0.019	4.9	0.012
Silver	0.0	0.000	0.0	0.000	0.0	0.000	0.0	0.001
Zinc	0.7	0.225	0.5	0.321	0.0	0.180	0.0	0.056

Source: Leland J. McCabe, "Problem of Trace Metals in Water Supplies: An Overview," in *Proceedings of the 16th Water Quality Conference*, University of Illinois, February 12–13, 1974.

tion informed us that 88 percent of the household water it has tested has exceeded EPA limits for some trace minerals.

Table 5-4: Limits for Minerals in Drinking Water

Mineral	Recommended Limit
Arsenic	0.05 mg/l[a]
Barium	1 mg/l[a]
Beryllium	no regulated limit[a]
Cadmium	0.01 mg/l[b]
Chromium	0.05 mg/l[a]
Cobalt	no regulated limit[a]
Copper	1 mg/l[b]
Lead	0.05 mg/l[a]
Manganese	0.05 mg/l[b]
Mercury	0.002 mg/l[a]
Molybdenum	no regulated limit[a]
Selenium	0.01 mg/l[a]
Silver	0.05 mg/l[a]
Tin	no regulated limit[a]
Vanadium	no regulated limit[a]

Source: This table is compiled by the author from information in *Drinking Water and Health* (Washington, DC: National Academy of Sciences, 1977).

a. *Based on EPA standard for drinking water quality listed in the National Interim Primary Drinking Water Regulations.*

b. *From USPHS Drinking Water Standards (1962). The EPA National Interim Primary Drinking Water Regulations supercedes the USPHS Drinking Water Standards, but does not regulate the limit of this mineral in drinking water.*

Hazardous Trace Minerals

Lead, one of the greatest threats to our health, has been warned against since before the Roman Empire. The ancients unintentionally poisoned themselves by drinking out of lead cups. Today industrial wastes, solid waste landfills containing paint and insecticides, and even home plumbing systems of lead and galvanized pipes bring the problem of lead contamination in drinking water dangerously close to home. And each day the devastating results of lead poisoning are becoming better known. The present safe level for lead is 50 micrograms per liter but the EPA has suggested that standard be reduced.

Researchers at the department of psychiatry, Downstate Medical Center, Brooklyn, New York, have found that the range of lead concentrations associated with mental retardation is much wider than previously assumed, and that equal amounts of lead can cause extremely variable toxicity.

Case in point: two groups of borderline or mildly retarded children were compared for blood lead levels. The group of children whose retardation was from "unknown" causes had significantly higher blood lead concentrations than the children whose retardation was from known causes, such as brain injury.[22] In other words, in cases of retardation where there is no apparent explanation, lead may be the culprit.

Oliver David, M.D., leader of the study, pointed out that "It is important to reiterate that no subject in this study had a history of lead poisoning as defined in present clinical practice. This report shows that lead at concentrations much lower than those causing the accepted clinical symptoms of lead poisoning may also be causing lead poisoning, albeit a type that is very differently defined."

A Scottish study attacked the same problem from a different angle. A host of researchers from two organizations, the University of Glasgow and the Glasgow Health Department, tested lead levels in the water of homes occupied during the first years of life by two groups of retarded and nonretarded children. *They found that the water lead levels were significantly higher in the homes of the retarded children, and that blood lead levels were also significantly higher in the retarded children.*

As in the American study, the retarded children were afflicted by "unknown causes." "As in all epidemiological studies of this type, it is not possible to prove that lead exposure causes mental retardation, but the results are strongly suggestive that this is so. There is good evidence that lead poisoning in young children may result in permanent impairment of mental capacity," report the researchers.[23]

"Slightly elevated" levels of lead in children's blood have also been associated with lowered scores on intelligence tests, according to Harvard psychiatrist Herbert Needleman, M.D.

The importance of the observation lies in the fact that the "high" blood levels of lead were well within the range previously considered safe.[24]

The insidious effects of lead do not stop with the nervous system. Scottish researchers found that, among men with high blood pressure, there were significantly more with high blood levels of lead than there were among men with normal blood pressure. Also, tap water samples from the homes of the men with high blood pressure had higher levels of lead than did the water taken from the homes of men with normal blood pressure.[25]

Low levels of lead can also slowly poison the body's ability to fight off infections and other harmful substances. Many experiments with animals have demonstrated that doses of lead small enough not to cause any obvious signs of poisoning severely hampered the ability to fight off bacterial and viral infections or survive the poisons bacteria produce.

Loren D. Koller, D.V.M., Ph.D., Associate Professor at the School of Veterinary Medicine, Oregon State University, and a colleague carried out experiments showing that lead hampers the immune response to disease by decreasing the number of cells in the body that produce antibodies. The overall effect is to reduce the number of antibodies prepared to fight off invading organisms, which results, as the Oregon researchers said, in "the increased mortality from bacterial and viral diseases in animals that are chronically exposed to lead."[26]

Fortunately, there is at least a partial antidote for lead—calcium. Experimental animals on low-calcium diets suffer the poisonous effects of lead much more than animals getting sufficient calcium. Plentiful calcium protects against lead even when it's injected into the animal's body.

So if you live in a hard-water area, some protection may be built into your water supply. If you don't, and lead has been detected in your water, your best bet (short of removing the lead altogether) is to supplement your diet with calcium—especially bone meal which contains phosphorus, another aid in removing lead from the body.

Cadmium is found in only small amounts naturally in water. But waste disposed from electroplating, photography, insecticide, and mining industries (cadmium is found in tandem with zinc) can increase levels dramatically. The most common source of cadmium in our drinking water, however, is from galvanized pipes and solder joints in copper plumbing.[27]

Cadmium is another highly toxic mineral. But what's really distressing about this trace mineral is that it accumulates in the body, and ten to twenty years of stockpiling cadmium can lead to a sudden explosion of poor health: high blood pressure, kidney disease, and emphysema have been linked to cadmium. The EPA limit is 10 micrograms per liter.

But, again, there may be some protection from the poisonous effects of cadmium in the calcium found in hard water.

A group of rats was fed a diet containing cadmium and low levels of calcium. Another group was also fed a diet containing cadmium, but with high levels of calcium. Compared with the first group, those getting extra calcium absorbed significantly less cadmium.[28]

Iron may also shield you against cadmium.

"Levels of dietary iron that exceed the normal requirement offer almost complete protection against cadmium toxicity in the growing rat," wrote Orville Levander, Ph.D.[29]

Interestingly enough, zinc, which often occurs together with cadmium in nature, is also antagonistic toward this toxic mineral.

In one study, scientists fed young quail a diet containing cadmium. When zinc was added to the diet, the level of cadmium in their tissues dropped.

"Zinc is an important element in preventing the accretion [accumulation] of low levels of cadmium similar to those present in the diet of man," wrote the scientists.[30]

Mercury is rarely found in water—at least naturally. However, with the increased use of mercury compounds in paints, in seeds to prevent mold, and in industry, mercury contamination is on an upswing in some areas of the world, most notably in Japan.[31] Studies have shown that in this country only about 4 percent of our water supplies contain mercury in concentrations of more than 1 microgram per liter. At that level, mercury is considered tolerable. But the increasing exposure may present a potential hazard. Mercury exerts its main toxic effects on the nervous system and kidneys where it can cause grave damage. A more probable danger exists when mercury enters our food chain. In fish, for example, harmful mercury compounds are stored and concentrated. Some fish have been examined and found to have concentrations of mercury that surpass acceptable limits for food—even though examinations of the water supply show that, generally, our water does not contain dangerous amounts of mercury.[32] The EPA has set the limit for mercury at 2 micrograms per liter.

Arsenic leaches into our water supply from solid waste landfills and the agricultural use of certain insecticides. As a result, arsenic is widely distributed throughout the United States but its concentrations tend to be small. The limit set by the National Interim Primary Drinking Water Regulations is 50 micrograms per liter. The problem is, waters may be contaminated with many different forms of arsenic, each of which carries its own array of hazards. Even worse, the more toxic compounds are retained by the body in greater amounts and are excreted more slowly than the less harmful ones.

Arsenic affects the tissues of the digestive tract, kidneys, liver, lungs, and skin. It is very damaging to capillaries, so that arsenic poisoning can result in gastrointestinal hemorrhage and heart abnormalities. There is also some evidence that high concentrations of arsenic in drinking water are associated with an increased incidence of skin cancer.[33]

Essential Trace Minerals

Remember, while it's best to avoid all traces of the minerals just mentioned, there are some minerals we can't live without—minerals which can be harmful in high doses but, when supplied in just the right amounts, can contribute to optimum health. Following is a list of some of the most common essential trace minerals.

Chromium concentrations in water are limited since it is not a very soluble mineral. And, of course, this is one mineral which most of us could use more of. Chromium, it seems, plays a vital role in our body's metabolism of sugar. It latches onto the body's own insulin and helps to make it more effective in its handling of sugar and carbohydrates. As such, many researchers—including Walter Mertz, M.D., chairman of the United States Department of Agriculture's Nutrition Institute and pioneer in chromium research—consider chromium the missing link in the prevention of late-onset diabetes.[34]

Chromium may also be a key to the prevention of coronary artery disease. In 1960, Dr. Henry J. Schroeder began an exhaustive study on the effects of small amounts of trace elements on rats and mice from weaning until natural death. Interestingly enough, the animals on chromium grew faster, survived longer, and, at death, had no cholesterol deposits in the arteries of their hearts. Another finding was that chromium-deficient old rats developed cataractlike opacities of the cornea of the eye. While these animal studies cannot be literally translated into human terms, Dr. Schroeder did find that, at autopsy, persons dying of coronary artery disease had no chromium in the arteries of their hearts, while persons who died accidently did.[35]

Cobalt likewise occurs in tiny amounts in natural waters. As an essential component of vitamin B_{12}, cobalt is important to nerve function and red blood cell formation.[36]

Copper is a minor constituent of natural waters. However, the copper found naturally is often supplemented with copper from corroded plumbing and industrial wastes, corrosion which increases when water is chlorinated, soft, or acidic. Also, as mentioned earlier, some water companies introduce copper into reservoirs to control the growth of algae. In trace amounts, copper is vital to bodily functions. For example, it helps in the formation of water from oxygen and hydrogen at body temperatures. In higher amounts, however, it can become dangerous to the human body, causing irritation of the gastrointestinal tract and possibly mental disorders.[37]

Iron is everywhere—in the earth's crust, in plants and animals, in sea water, and, of course, in drinking water. And, it's a good thing because iron is an integral component of red blood cells and is involved in the transport of oxygen to all body systems.

A lack of iron—a condition called iron-deficiency anemia—means that every system in your body is gasping for oxygen. The possible symptoms of anemia read like an excerpt from a hypochondriac's diary: Fatigue. Heartburn. Dizziness. Brittle fingernails. Hair loss. Irritability. Overall itching. Nausea after meals. Headaches. Sore tongue. Loss of appetite. Weak legs. Pale skin.

Obviously, we could all do with more of this trace mineral. Unfortunately, though, cadmium and lead—two hazardous trace minerals—can cut our body's absorption of this one essential mineral.

Lithium generally finds its way to our water supply through natural channels. It's considered a mood stabilizer and with good reason. Studies in Texas found lower cardiovascular mortality rates, a smaller number of admissions for mental disorders, and a lower frequency of violent behavior such as homicides and suicides in certain areas than in matched neighboring areas that differed in only one respect—the water in the low-rate areas contained much more lithium.

Similarly among the Pima Indians of Arizona, the low rate of coronary heart disease and of stomach ulcers (both stress-related diseases) has been linked to the high lithium content of their water supplies (100 micrograms per liter), as compared with the water in the rest of the United States (which averages about 2 micrograms per liter).[38]

Manganese is less abundant in water than iron. Nevertheless, it is required by all living things, taking part in a number of reactions involving enzymes. It's also reassuring to know that, even in large amounts, its potential for human harm is slim. The limit for manganese is 0.05 milligram per liter.[39]

Molybdenum is frequently used in metallurgy and is often a constituent of fertilizers. It has been found in surface and groundwaters at very low concentrations. Although it aids the body in its production of uric acid, excessive amounts have been associated with gout and bone disease.[40]

Nickel usually exists in water as the result of human activities. It is probably essential to human nutrition, and an excess of nickel is excreted in the feces. No limits have been set for its concentration in drinking water.[41]

Selenium in drinking water varies widely, depending on the selenium content of soil and rocks. The present limit is 10 micrograms per liter. But evidence suggests that the people who get a significant dose in their food and water could be the lucky ones. Studies around the world have shown that, as selenium intake approaches 300 micrograms a day, protection against certain types of cancers seems to be enhanced. One of the most striking examples of this is breast cancer. Gerhard N. Schrauzer, Ph.D., a selenium researcher at the University of California in San Diego, analyzed data from seventeen countries and found that, as levels of selenium in the blood rise, breast cancer death rates fall.

In addition, Dr. Schrauzer took specially bred mice that normally develop spontaneous mammary tumors and fed them selenium in their drinking water. Instead of the usual cancer incidence of 80 to 100 percent, only 10 percent developed tumors.

Selenium also seems to be instrumental in preventing heart disease. Raymond J. Shamberger, Ph.D., and associates at the Cleveland Clinic Foundation noted in a paper presented at the 12th annual University of Missouri trace element conference in 1978, rats and lambs fed selenium-deficient diets have abnormal electrocardiograms and blood pressure changes. And humans are apparently affected as well. When the researchers compared mineral intakes with heart and artery disease death rates in twenty-five countries, they uncovered a significant link: where selenium consumption was lowest, deaths due to coronary disease that affects the arteries leading into the heart were greatest.

They found one other fascinating correlation. Coronary artery disease also tended to rise where cadmium intake was high. Cadmium is a mineral pollutant that competes with selenium for biochemical binding sites within the body. Cadmium is a suspected cause of high blood pressure.

It's interesting to note that selenium, when present in large enough amounts, seems able to latch onto cadmium and divert it from some of its destructive missions inside the body. A recent University of Cincinnati College of Medicine study, for example, shows that injected selenium blocks the usual destructive effects of cadmium on the testicles of laboratory animals. The researchers speculate that, thanks to selenium, the cadmium (at least in the testicles) is converted into "a nontoxic or inactivated form."[42]

Tin is seldom found in water. It is extremely difficult to test for. The major source of tin in man's diet is canned food and drinks. It may be essential to human nutrition. No maximum contaminant level has been set.[43]

Vanadium in water is generally the result of mining. Recently, it has been introduced into the environment in large quantities. Vanadium is probably essential to human nutrition, and is thought to be protective against atherosclerosis. The limit for vanadium concentration in drinking water is 1 milligram per liter.[44,45]

Zinc is found in water in varying concentrations—anywhere from 2 to 1,200 micrograms in a liter. The EPA limit is 5 milligrams per liter.[46] Its presence in large quantities is the result of urban and industrial runoff. In small, trace amounts, the way it occurs naturally in the soil and water, zinc can be a savior to your system. In fact, perhaps more research is being published on the amazing and versatile powers of zinc than on any other trace mineral.

It's easy to see why. Zinc is found in every tissue of the body and plays a key role in the health of every bodily system.

In dramatic detail, study after study show zinc's importance to physical health and mental well-being. Let's take a look at some recent and exciting research.

One of zinc's most remarkable feats is healing a rare, inherited disorder, acrodermatitis enteropathica (AE). AE preys on young children, ripping open their skin with horrible, festering lesions and stunting their growth. Its only cure was death.

But in 1974, researchers discovered that zinc supplements cure the disease, completely clearing up the severe physical disfigurement—as long as patients continue to take zinc. Recently, researchers also discovered that zinc supplements cause amazing changes in the *mental* condition of AE patients.

One common eye disorder that can be caused by zinc deficiency is night blindness, the inability of the eyes to adapt to darkness. In a study of six men with night blindness, all six regained normal vision after taking zinc supplements for two weeks.[47]

Giving zinc to acne patients, a Swedish researcher found that the mineral cleared up 85 percent of their pimples after twelve weeks.[48]

And the list of zinc's relief-giving powers goes on and on. Studies have shown zinc can protect the body against the environmental pollutants lead and cadmium. It can also speed the healing of stomach ulcers, improve the flexibility of rheumatoid arthritis, and reduce a swollen prostate gland—a problem that afflicts almost all men over sixty.

While it's possible that industrial pollution can contribute to excessive levels of zinc in your drinking water—and that drinking zinc in quantities of 40 to 50 milligrams can make you sick—in most cases, we're getting too *little* zinc, not too much.[49]

"Many Americans may be consuming only marginally adequate levels of zinc," said a study on the zinc intake of elderly persons. And, "consumption of marginal levels [of zinc] over a span of 60 years or more could account for some of the degenerative changes associated with aging."[50]

Besides these, there are a few minerals which do not fit either category of hazardous or essential. Based on what we know at the present time, they are neither extremely harmful (at least not in trace amounts) nor healthful. According to the evidence that's been filtering in, however, it would be wise to avoid them as they can be dangerous in excessive amounts.

Barium, for one, can enter water supplies through industrial waste discharges. In small doses, we are told, barium does no harm. In fact, it is frequently used as a diagnostic tool; when taken internally it serves as a contrast medium for x-rays. There is some evidence accumulating, however, which casts a shadow over the safety of barium. Researchers at the University of Illinois at the Medical Center in Chicago have recorded higher death rates due to heart disease and all cardiovascular diseases and higher blood pressure rates among male residents of Illinois communities with elevated amounts of barium in their drinking water.[51]

Beryllium generally is used in the manufacture of metal alloys. It is harmless in small amounts, but researchers have found some association between beryllium and cancer in laboratory animals. The EPA has not set any limit on beryllium in drinking water.[52]

Silver is sometimes added to municipal water supplies to act as a disinfectant. It is also used in certain types of activated carbon water filters, to prevent or slow the growth of bacteria. We are told that it should pose no problem. And in trace amounts it probably doesn't. Large doses, however, can cause anemia and possibly death.[53] Occupational and medical exposure to silver causes a permanent ashen-grey discoloration of the skin, internal organs, and membranes lining the inner surface of the eye lids.[54]

Knowing which minerals are present in your home water supply could help you make some important decisions about your health. Like considering plumbing remedies or looking for an alternative water source if your water is high in the hazardous minerals. Or supplementing your diet with extra doses of the essential minerals you might be missing out on in your water.

But, before you do that, it's important that you understand the limitations of water testing. Water tests tell you how much of which minerals you're getting in your water. But it doesn't tell you how much you might be picking up from the air you breathe or the food you eat —two prime sources of trace minerals. In other words, it won't give you a total picture of trace minerals in your body. One way of getting a clearer overall view of trace minerals on your health is by having your hair tested.

Hair testing was first developed to catch criminals. Hair found at the scene of a crime was analyzed in the hope that it would help to establish the identification of a criminal.[55] While hair testing never really became firmly established in forensic medicine, the time spent analyzing hair paid off in another way. Scientists found that hair serves as a record of metabolic disorders, poisoning, and exposure to minerals.

A single strand of hair can carry a record of these physical events for as long as seven years. Moreover, some studies have shown that concentrations of metals in the hair shaft can reflect the amounts of those same metals stored within the organs of the body.[56]

Most hair studies have been aimed at finding ways to enhance the diet of people who are sick, mentally or physically, and even as a way to help slow learners. Researchers have found that people with certain kinds of diseases have abnormal amounts of some minerals in their hair. For example, a firm relationship has been found between lead levels in hair and lead poisoning. And when people consume mercury, nickel, or cadmium, it also shows up in their hair.[57] Researchers at McGill University tested the hair samples of a group of children and could tell, with 98 percent accuracy, which children had learning problems. Those kids had higher amounts of lead, cadmium, and manganese in their hair, but lower than normal amounts of lithium, chromium, and zinc.[58]

But nothing is without some problems—including hair testing. To date, there has been no uniform testing procedure. As a result, one individual's hair could show a high concentration of lead in one test, but in a second test done the same day, could show a low lead concentration. This kind of discrepancy results from testing different portions of a length of hair. It's known that the end of the hair closest to the scalp usually has smaller concentrations of minerals, perhaps because it is younger and has picked up less from the environment.

A second problem is that the hair sample may be prepared for testing in several ways. One scientist may clean the sample with ether, while another uses alcohol, resulting in different readings of mineral concentrations. A third variant is the test procedure itself. Methods of analysis include neutron activation, photon activation, atomic absorption spectrometry, and particle-induced x-ray emission analysis.[59]

For testing, the Soil and Health Foundation requests about a tablespoon of hair taken from the nape of the neck three days after a shampoo. It is tested by using atomic absorption spectrometry.

Studies have shown that deficiencies of calcium, magnesium, cobalt, iron, manganese, and zinc are fairly common, but the mineral composition of hair may be unique for each individual. It seems to be affected by age, sex, and season.

Only a few studies have been done to determine the correlation of minerals in drinking water with minerals in hair. One test analyzed the hair of residents of Yellowknife, in the Northwest Territories of Canada. These samples showed high levels of arsenic. Another test confirmed that the drinking water in Yellowknife also had high arsenic levels.[60]

The most extensive study of mineral levels in hair and drinking water was done on forty-three people who live in the little town of Milan, New Mexico. The home wells in Milan yielded water containing substantial amounts of selenium—anywhere from a norm of 10 micrograms per liter to 3,900 micrograms per liter. The source of this selenium was presumed to be a nearby disposal site of uranium mill tailings.

Scientists tested samples of each individual's tap water, along with samples of his urine, whole blood, and hair. As the level of selenium rose in the drinking water, it also rose in the samples of hair and urine. In other words, people who drank from wells containing the highest levels of selenium also had the highest levels in their urine and hair, but not in their blood. For some reason, there seems to be an upper limit for selenium in blood, and increased selenium consumption is not reflected in a blood test once that upper limit has been reached. Unfortunately, blood tests are widely used, while hair tests are not.[61]

You and members of your family may wish to have your hair tested for mineral concentrations. These tests are extremely beneficial. To begin with, you may find that certain individuals in your family have high levels of copper or lead, while others do not. If the copper and lead is coming from the drinking water supply, it may show up only in those family members who are at home most of the time, while those who drink part of their daily water at school or work may have lower concentrations.

Hair analysis shows the entire body burden. For example, if one member of the family is exposed to copper in his work environment, and also exposed to copper in his drinking water, the total will show up in his hair sample. For that individual, even the smallest amount of copper in the water he drinks might be harmful to his health, while other family members may be able to tolerate higher levels.

One of the problems the National Academy of Sciences had in establishing safe levels for metals in drinking water is that nobody knows the total exposure an average person gets from food, air, and water. Hair analysis is a good way to estimate total exposure, and then assess how much of any metal you are willing to drink in your water.

Testing for Organic Chemicals

In a previous chapter, we discussed the presence of organic chemical compounds in drinking water. We defined them as being compounds based on carbon, and which can be either natural or synthetic in origin.

There are several ways these organics can get into our water. They can come from natural sources, from water treatment, from industry, from sewage treatment plants, from runoff, and from spills and accidents.

Perhaps the most disturbing type of contamination is the kind that results from the water purification process itself. You may remember that, when chlorine interacts with natural humic acid, or with algae, or with other organic material in water, halogenated organic chemicals are created. They are found in virtually all treated waters, and many have been linked to cancer.[62]

Technically, these dangerous compounds are known as halo-organics, and the type of halo-organic most likely to show up in your water is a group called trihalomethanes (THMs). The most common of the THMs found in drinking water is chloroform.

Even though the Food and Drug Administration banned the use of chloroform in foods and drugs, it still shows up because it is made right in the water as chlorine and humic acids interact. Investigators have found chloroform in water at concentrations ranging from 0.001 milligram per liter to 2.0 milligrams per liter in Durham, North Carolina. Trihalomethanes other than chloroform have been detected in potential concentrations as high as 0.784 milligram per liter.[63] In fact, more than 700 specific organic chemicals have been identified in drinking water supplies in this country. Concentrations range from almost nothing in protected groundwaters to frighteningly high levels in some surface waters and contaminated groundwaters.

Yet, many water authorities feel that the organic compounds that have been identified so far may be just a small fraction of the total organic chemical content. And—more than just possible, it is probable —these unidentified substances may be even more toxic than the ones already known.[64]

The Environmental Protection Agency has set limits for organics in our water. But the limits themselves are very limited. The maximum

(continued on page 112)

Table 5-5: Organic Compounds Present after Chlorination (measured in micrograms per liter)

Water Districts by State		Chloroform		Bromodichloromethane	
Arizona—	Phoenix	R[a]	0.2	R	*
		T	9	T	15
	Tucson	R	0.1	R	*
		T	0.2	T	0.8
California—	Coalinga	R	0.2	R	*
		T	16	T	17
	Concord	R	0.3	R	0.3
		T	31	T	18
	Dos Palos	R	*	R	*
		T	61	T	53
	Los Angeles	R	0.1	R	*
		T	32	T	6
	San Diego	R	*	R	*
		T	52		30
	San Francisco	R	*	R	*
		T	41	T	15
Colorado—	Denver	R	0.2	R	*
		T	14	T	10
	Pueblo	R	0.2	R	*
		T.	2	T	2
Delaware—	Claymont	R	0.3	R	0.4
		T	23	T	11
Florida—	Jacksonville	R	*	R	*
		T	9	T	4

Table 5-5—*continued*

Water Districts by State	Chloroform		Bromodichloromethane	
Florida— Miami-Dade	R	*	R	*
	T	311	T	78
Indiana—Indianapolis	R	0.1	R	*
	T	31	T	8
Iowa— Clarinda	R	0.2	R	*
	T	48	T	19
Davenport	R	0.4	R	*
	T	88	T	8
Kansas—Topeka	R	0.4	R	0.8
	T	88	T	38
Maryland—Baltimore	R	*	R	*
	T	32	T	11
Massachusetts—Boston	R	*	R	*
	T	4	T	0.8
Michigan—Detroit	R	0.2	R	*
	T	12	T	9
Minnesota—St. Paul	R	0.2	R	*
	T	44	T	7
Missouri— Cape Giradeau	R	0.2	R	*
	T	116	T	21
Kansas City	R	*	R	*
	T	24	T	8
St. Louis	R	*	R	*
	T	55	T	13
New Jersey— Little Falls	R	0.3	R	*
	T	59	T	16
Toms River	R	0.4	R	*
	T	0.6	T	0.8
New Mexico—Albuquerque	R	*	R	*
	T	0.4	T	1
New York— Buffalo	R	*	R	*
	T	10	T	10
New York City	R	*	R	*
	T	22	T	7
North Dakota—Grand Forks	R	*	R	*
	T	3	T	1
Ohio— Cincinnati	R	0.5	R	*
	T	45	T	13
Cleveland	R	*	R	*
	T	18	T	9

Table 5-5—*continued*

Water Districts by State	Chloroform		Bromodichloromethane	
Ohio— Columbus	R	0.1	R	*
	T	134	T	8
Dayton	R	*	R	*
	T	8	T	8
Youngstown	R	*	R	*
	T	80	T	5
Oklahoma—Oklahoma City	R	*	R	*
	T	44	T	28
Pennsylvania— Philadelphia	R	0.2	R	*
	T	86	T	9
Western Pa. Water Co.	R	0.3	R	*
	T	8	T	2
Rhode Island—Newport	R	*	R	*
	T	103	T	42
South Carolina—Charleston	R	0.2	R	*
	T	195	T	9
Tennessee— Chattanooga	R	0.9	R	*
	T	30	T	9
Memphis	R	0.2	R	*
	T	0.9	T	2
Texas— Brownsville	R	*	R	*
	T	12	T	37
Dallas	R	0.1	R	*
	T	18	T	4
San Antonio	R	*	R	*
	T	0.2	T	0.9
Utah—Salt Lake City	R	0.2	R	*
	T	20	T	14
Washington—Seattle	R	0.2	R	*
	T	15	T	0.9
West Virginia—Wheeling	R	0.2	R	*
	T	72	T	28
Wisconsin— Milwaukee	R	0.2	R	*
	T	9	T	7
Oshkosh	R	*	R	*
	T	26	T	4

Source: M. T. Gillies, ed., *Drinking Water Detoxification* (Park Ridge, NJ: Noyes Data Corporation, 1978), pp. 114–119.

a. R = raw water; T = treated water; * = substance not found in water.

contaminant level set for all THMs, including chloroform, is 0.10 milligram per liter. But these standards are initially applicable only to water systems serving 75,000 or more people, and will not go into effect until, perhaps, the mid-1980s. Additionally, the maximum of 0.10 milligram per liter was chosen because practical considerations were taken into account—including the feasibility of achievement.

Once testing becomes standard, scientists who measure THM levels in our water will be able to use that information to determine the presence of other, unidentified halogenated compounds by treating THMs as an indicator—in much the same way coliform bacteria are used as an indicator of other bacteriological contaminants.

The tests performed for THMs and other organic chemicals are extremely sophisticated. In fact, current technology may be inadequate for the task. For example, in 1974 and 1975, a survey was conducted to see what organic compounds were in our water, and in what concentrations. This massive testing was called the National Organics Reconnaissance Survey (NORS) and found 129 organic compounds in finished water.[65] Yet, the tests done at that time failed to detect several important classes (not single types) of carcinogens because the technology for analysis had not yet been developed.

If you want to know whether organic compounds exist in your drinking water, the only way to find out is to have your water tested. But it will cost you an arm and a leg. A single test for just the THMs will cost anywhere from $120 to $250, and will cover only a few of the 700 or so organics identified in tap water. Very expensive precision equipment is critically important when measuring such small amounts of a substance. In some cases, it must be capable of measuring one single part of a billion parts. Consequently, costs are quite high. In fact, cost is one big reason why neither government nor private agencies have launched any massive programs to detect all the organics in water.

Instead of reaching into your pocket to pay someone to test your water, first use your own common sense. Find out where your water comes from; ask your water authority (or the Department of Environmental Resources) about the quality of the raw water, and then find out how that water is treated and distributed. With this information you can figure what probable organic contaminants are in your water with a fair amount of accuracy. While you will not know the levels at which these chemicals are present, you will know which are in your drinking water.

If You Have Public Water:

Virtually all public water supplies are chlorinated to prevent the spread of waterborne diseases. And, virtually all chlorinated supplies contain some organic contaminants as a result of that treatment. Moreover, your water may contain additional organics from other sources.

When the National Organic Reconnaissance Survey studied the drinking water of eighty cities in 1975, chloroform was found in all samples, regardless of whether the water was taken from a clean or polluted source. Drinking water can contribute anywhere from 0 to 90 percent or more of a person's total dietary intake of chloroform.[66] Other sources of chloroform are food and air. The survey revealed that other organic chemicals found frequently in water included: dibromochloromethane, which can cause genetic mutations but has never been tested for carcinogenicity; carbon tetrachloride; 1,2-dichloroethane; vinyl chloride; and pesticides like dieldrin and lindane. Thousands of these compounds may be present in our drinking water in very small amounts—raising the risk of cancer probability.

If your treated water comes from a lake, river, stream, or other surface body, there is an even greater probability that organic chemicals may be present in your water, and at higher concentrations than if your water source comes from the ground. There are, however, notable exceptions. The reason surface waters are prone to higher concentrations of organics is that these waters contain more humic acids and other precursors to the formation of organics. Because many surface waters are used for waste disposal, the potential for pollution by organics is greater. And in highly industrialized areas, dangerous organic compounds can be present as the result of poor disposal procedures.

Some of the latest water surveys reveal the presence of a previously undetected contaminant called polynuclear aromatic hydrocarbons. The contamination is widespread, and results from combustion of fossil fuel. Auto exhaust washed from the streets into water supplies, crankcase oil, fuel mixtures, and the like are sources of this new class of pollutants. Generally, surface waters are more affected by this form of pollution than groundwaters.

If you are supplied with public drinking water and you are unsure of whether it is chlorinated, call your municipal water authority or your department of health.

If You Provide Your Own Water:

Determining whether organic compounds are in your water is extremely difficult unless you pay to have a sample tested. In the case of private supplies, educated guesswork is not very educated.

But this much is almost certain. If your drinking water comes from a surface source, there is a greater probability that organics are present in it. Surface supplies are open to industrial and agricultural pollution. If your water source is a well or a spring, and if the area around you is pristine and virginal—not farmed, industrialized, or built up into housing developments—your water may be uncontaminated by organic chemicals. But the only way to find out for sure is to have it tested.

If you live in a city, the chances are greater that some form of natural or synthetic organic compound may have filtered down into the water table. If you live in farmland, the chance is good that some organic chemical originating from pesticides or herbicides may have found its way into your source. Again, the only way to know is to have your water tested by a sophisticated laboratory.

Radiation Testing

Minute traces of radioactivity are found in just about all water. The concentration and composition of these radioactive particles vary from place to place, depending on the soil and rock strata of an area. Most of this radiation is natural, and comes from the earth's crust, from cosmic rays, and from the earth's atmosphere.[67]

The development of nuclear weapons and nuclear power plants has increased the possibility of radioactive contamination. When the Three Mile Island Nuclear Power Plant was disabled on March 28, 1979, an internal accident allowed water to become contaminated by highly radioactive matter. Some hot water was discharged into the Susquehanna River after the "event," creating the possibility of radioactive contamination to anyone downriver who might use that water for drinking.

We know that nuclear weapons can contaminate drinking water because of the history of water wells on the island of Bikini, which was used as a hydrogen bomb test site in 1954. The residents, who returned to their homes in 1969 after fifteen years of displacement, found that levels of strontium-90 in their well water exceeded standards. Also, the breadfruit, bananas, and coconuts the islanders use as staples also contained excessive radioactivity. In 1978, Congress asked for $15 million so that these people again could be relocated.[68]

In this country, a survey of groundwater revealed that one-quarter of the 507 samples taken had elevated levels of radon. The samples highest in radon came from Maine and New Hampshire, where about 80 percent were above safe levels.[69]

NOTES

1. Henry A. Schroeder, M.D., *Trace Elements and Man: Some Positive and Negative Aspects* (Old Greenwich, CT: Devon-Adair Co., 1973), p. 13.

2. Ibid., p. 31.

3. Ibid., preface, p. viii.

4. Ibid., p. 154.

5. Ibid., p. 92.

6. "How Trace Elements in Water Contribute to Health," *World Health Organization Chronicle,* October, 1978, p. 382.

7. Schroeder, *Trace Elements and Man,* p. 55.

8. *World Health Organization Chronicle,* p. 382.

9. Ibid., p. 383.

10. Ibid., p. 382.

11. *Drinking Water and Health* (Washington, DC: National Academy of Sciences, 1977), p. 205.

12. Ibid., pp. 205, 207.

13. Stephen C. James, "Metals in Municipal Landfill Leachate and Their Health Effects," *American Journal of Public Health,* May, 1977, pp. 429–430.

14. *Drinking Water and Health,* p. 206.

15. Leland J. McCabe, "Problem of Trace Metals in Water Supplies: An Overview," *Proceedings of the 16th Water Quality Conference,* University of Illinois, February 12–13, 1974, p. 6.

16. *Drinking Water and Health,* pp. 210–211.

17. Leland J. McCabe et al., "Survey of Community Water Supply Systems," *Journal of the American Water Works Association,* November, 1970, pp. 670–687.

18. McCabe, *Proceedings of the 16th Water Quality Conference,* p. 7.

19. "The Water Test," *Soil and Health News,* August, 1979, p. 7.

20. Ibid.

21. Ibid.

22. Oliver David et al., "Low Lead Levels and Mental Retardation," *Lancet,* December 25, 1976, pp. 1376–1379.

23. A. D. Beattie et al., "Role of Chronic Low-Level Lead Exposure in the Etiology of Mental Retardation," *Lancet,* March 15, 1975, pp. 589–592.

24. B. D. Colen, "Study Links Level of Lead in Blood to IQ Test Performance," *Washington Post,* May 6, 1978.

25. D. G. Beevers et al., "Blood-Lead and Hypertension," *Lancet,* July 3, 1976, pp. 1–4.

26. Loren D. Koller and Sharlotte Kovacic, "Decreased Antibody Formation in Mice Exposed to Lead," *Nature,* July 12, 1974, pp. 148–149.

27. *Drinking Water and Health,* pp. 236–237.

28. L. B. Sasser et al., "The Effect of Calcium and Phosphorus on the Absorption and Toxicity of Cadmium," *3rd International Symposium on Trace Element Metabolism in Man and Animals,* July 25–29, 1977.

29. Orville A. Levander, "Nutritional Factors in Relation to Heavy Metal Toxicants," *Federation Proceedings,* April, 1977, pp. 1683–1687.

30. R. M. Jacobs et al., "Cadmium Metabolism: Individual Effects of Zinc, Copper, and Manganese," *Federation Proceedings,* March, 1977, p. 1152.

31. Schroeder, *Trace Elements and Man,* p. 91.

32. *Drinking Water and Health,* pp. 271, 272, 274, 275.

33. Ibid., pp. 316, 319, 322, 326.

34. Schroeder, *Trace Elements and Man,* p. 71.

35. Ibid., pp. 71–73.

36. *Drinking Water and Health,* p. 247.

37. Ibid., pp. 250–252.

38. *World Health Organization Chronicle,* pp. 383–384.

39. *Drinking Water and Health,* pp. 265, 267.

40. Ibid., pp. 279, 284.

41. Ibid., pp. 285, 287.

42. Robert J. Niewenhuis and Peggy L. Fende, "The Protective Effect of Selenium on Cadmium-Induced Injury to Normal and Cryptorchid Testes in the Rat," *Biology of Reproduction,* August, 1978, pp. 1–6.

43. *Drinking Water and Health,* pp. 292, 294.

44. Ibid., pp. 296–297.

45. M. T. Gillies, ed., *Drinking Water Detoxification* (Park Ridge, NJ: Noyes Data Corporation, 1978), p. 21.

46. Ibid.

47. Stanley A. Morrison et al., "Zinc Deficiency: A Cause of Abnormal Dark Adaptation in Cirrhotics," *American Journal of Clinical Nutrition,* February, 1978, pp. 276–281.

48. Gerd Michaelsson, Lennart Juhlin, and Anders Vahlquist, "Effects of Oral Zinc and Vitamin A in Acne," *Archives of Dermatology,* January, 1977, p. 31.

49. Gillies, *Drinking Water Detoxification,* p. 21.

50. Janet L. Greger and Brenda S. Sciscoe, "Zinc Nutriture of Elderly Participants in an Urban Feeding Program," *Journal of the American Dietetic Association,* January, 1977, pp. 37–41.

51. "Higher Blood Pressure Rates and Death Rates Due to Cardiovascular Diseases Recorded in Communities with High Levels of Barium in Drinking Water," *News: University of Illinois at the Medical Center,* March 29, 1979.

52. *Drinking Water and Health,* pp. 232–235.

53. Gillies, *Drinking Water Detoxification,* p. 20.

54. *Drinking Water and Health,* p. 291.

55. "Hair and Health," *Chemistry,* March, 1979, p. 28.

56. "Diagnostic Clues from Hair 'Biopsy': Disorders of Hair Growth and Structure Can Signal Deeper Pathology," *Medical World News,* March 24, 1975, pp. 67–69.

57. Sidney A. Katz, "The Use of Hair as a Biopsy Material for Trace Elements in the Body," *American Laboratory,* February, 1979, pp. 44–52.

58. *Chemistry,* p. 28.

59. Katz, *American Laboratory,* pp. 44, 48.

60. Ibid., p. 44.

61. Jane L. Valentine, Han K. Kang, and Gary H. Spivey, "Selenium Levels in Human Blood, Urine, and Hair in Response to Exposure via Drinking Water," *Environmental Research,* December, 1978, pp. 347–355.

62. Robert L. Jolley, Hend Gorchev, and D. Heyward Hamilton, Jr., eds., *Water Chlorination: Environmental Impact and Health Effects,* Volume 2 (Ann Arbor, MI: Ann Arbor Science, 1978), p. 529.

63. Gillies, *Drinking Water Detoxification,* p. 108.

64. Ibid., p. 48.

65. Ibid.

66. Ibid., pp. 49, 50.

67. *Drinking Water and Health,* p. 858.

68. Joseph H. Highland et al., *Malignant Neglect* (New York: Alfred A. Knopf, 1979), p. 160.

69. Ibid., p. 161.

6 Let's Clean It Up

You *can* clean up your own water supply. Even if you do not want to invest in any manufactured water filters or purifiers, there are still measures you can take to make sure that you have pure and healthful water to drink.

This chapter will discuss the simplest ways to supply yourself with pure water—from buying domestic and imported bottled waters to trying some Fast Tricks on the water that comes from your faucet.

The Quick Fix—Bottled Water

Whether Perrier or Poland Water, Hungary's Kristalyviz or Canada's Dry, people are swigging more bottled water than ever before. One Wall Street investment firm considers it the biggest potential growth industry of the eighties. Their judgment is based on fact. In 1979, sparkling waters were the fastest growing sector of the United States's $12 million soft drink industry. Since 1976, sales of bottled water have grown by 10 percent annually. Industry experts say the growth is based on the increased health-consciousness of Americans, but sales also have been aided by aggressive advertising campaigns.[1]

While sales are just beginning to bubble in the States, bottled waters have been a traditional favorite of Europeans. Spas have long

120

been visited by people hoping to cure their ills. Most of them have bottled their waters since the turn of the century. While many of the spas' so-called cures are probably the result of wishful thinking, some may actually result from a specific ingredient inherent in the water. For instance, in England there is a legend about a poor, leprous swineherd named Bladud, who was cured of his disease when he lay down with his swines in their wallow. That wallow, it turns out, is the site of Bath, famous for its hot sulfurous springs. And modern medicine has indeed shown that sulfur can have a beneficial effect on leprosy, giving the legend a small foothold in truth.[2]

In this country, both George Washington and Secretariat have sipped the waters at Saratoga. But American bottlers of water don't make any health claims about their products. Rather they emphasize good taste and purity.

Several types of domestic bottled water are offered to the public:

Drinking water may come from a spring or well, or may be processed and purified.

Distilled drinking water is produced by changing the water to a vapor and allowing it to condense into a liquid form. This kind of bottled water has no solids, minerals, or trace elements. Most people think it tastes pretty flat, and mainly use it in their steam irons.

Fluoridated water is bottled drinking water to which fluoride has been added in controlled amounts.

Purified water is water that has been processed, either by distillation, deionization, reverse osmosis, or electrodialysis. The technique used isn't very important. What is, is that the majority of minerals have been removed, so that none remains in quantities greater than 10 parts per million.[3]

More than 85 percent of all bottled water is sold in five states—California, Florida, Illinois, New York, and Texas, with California consuming more than the rest by far.[4] In fact, Californians have access to a type of vending machine that purifies tap water for only a few cents. Consumers just bring along a jug of their own water, pour it into the unit, wait for the machine to do its work, and receive the cleansed water back in their jug.[5]

Californians use bottled water for a variety of reasons, not the least of which is that it's a trendy thing to do. Additionally, during dry spells, their tap water increasingly comes from the Colorado, a mineral-laden river that produces a mighty salty drink. Another problem many Californians have to contend with is that the tap water is warm. Water mains are set just below the ground—not three or four feet below, as required in climates where the ground freezes—and so water arrives at the tap unappetizingly lukewarm, as well as salty.

Bottled water sales also spurted in Florida after a typhoid fever epidemic in Miami, and in Cincinnati, after carbon tetrachloride was accidentally spilled into the Ohio River.

At one time, bottled water was considered a "convenience service," providing cool water to offices, schools, and other public buildings. Usually, it was delivered in 5-gallon carboys. While this service still exists, the latest rise in consumption is based on the purchase of quart or gallon bottles, available in grocery stores and supermarkets.

Sophisticates and dieters are riding the crest of bottled water sales, ordering it by brand name in bars and restaurants in place of wine or liquor. At Regine's, the elite Manhattan disco, a glass of Perrier, water imported from France, costs $6. Mae West prefers to drink Poland Water, and Gloria Swanson has been known to bathe in it.[6] But the less famous generally settle for a local brand of bottled water, bought at the market or delivered to their homes.

If your tap water is unsatisfactory, bottled water may be an interim solution to your problems, or your final answer. If, for example, you plan to buy a purification device, you may want to use bottled water only until the unit is installed. People who work in a spot where tap water is pure, or where bottled water is provided by their employer, may find bottled water convenient for their limited needs at home, and a pennywise alternative to purification devices.

Despite the many advertising claims about absolute purity, bottled water is not without some problems. When the Environmental Protection Agency studied bottled waters a few years ago, they found some samples containing fecal coliforms and other intestinal bacteria. In fact, when the EPA researchers allowed this water to set for two months, they found the bacteria grew and multiplied to the point where they were "too numbrous" to count.[7]

However, bottlers attempt to ensure the purity of their product. Ninety percent of the country's bottled water companies belong to the American Bottled Water Association, and adhere to its purity standards.[8] In fact, the ABWA worked to bring about federal regulations for bottled water, coordinating regulations that previously had varied from state to state.[9]

Even with federal regulation, the quality of water can vary from bottler to bottler, depending on the water's source. For instance, some waters come from deep wells, some from natural springs, some are naturally carbonated, and others are just purified tap water.[10]

If the water is treated, generally it is filtered to remove all traces of suspended solids. Then it may undergo ion exchange to remove minerals, reverse osmosis for even more complete mineral removal, and disinfection. Many bottlers disinfect with ozone, a bactericidal gas, rather than chlorine because it doesn't leave behind any taste or odor. Others use chlorine to disinfect, then remove all traces of the chemical by filtering the water through granulated activated carbon (GAC).[11] However, most bottlers do not pass their water through GAC, and for that reason, traces of chlorine, chloroform, or chemicals present in the raw water may still be in the final product.

Mineral Waters

Of course, some bottlers don't treat the water at all. The fortunate few with access to natural mineral springs or artesian wells usually bottle right at the source. One of the oldest and best known is Mountain Valley Water, a natural spring water bottled at its source in Hot Springs, Arkansas. It is the only American bottled water distributed nationally. Foreign sources of pure mineral waters bottled at their source include Perrier, Vittel, Evian, Apollinaris, Fiuggi, and Spa Reine.

Most imported bottled water is high in mineral content and considered very flavorful. There are two ways to get mineral water—one way is to find a source that is naturally high in elements like calcium and magnesium, the other way is to add minerals to bland water. If a bottler does add minerals, he has to state it on the label. The kinds of mineral blends used are often key to the water's flavor, and as a result they are highly guarded company secrets.

Spring Water vs. Treated Water

There are other labeling restrictions, as well. For instance, only water from a natural spring can be called spring water. Other water cannot be designated natural spring water. Nor are bottlers allowed to use deceptive phrases like "spring-like," or "spring-fresh." If the label doesn't say it is spring water, then it isn't.

The same kind of ruling applies to the term *effervescence.* Only naturally bubbling water can be described as effervescent. Waters that have been pepped up with carbon dioxide after leaving the spring are not effervescent.

While bubbling mineral water may be a very trendy drink, it probably will not be your choice for day-to-day living. More likely you will choose domestic, nonbubbling water for cooking and drinking.

An Informal Taste Test

In an informal taste test at Rodale Press, editors were offered a selection of eight bottled waters, some sparkling and some still, to see how the famous and not-so-famous brands ranked when they were offered in unmarked glasses. The sparkling waters included: Apollinaris Water Bad Neuenahr (Germany), Perrier (France), Saratoga Vichy (New York), Canada Dry (national brand), and San Pellegrino (Italy). The still waters offered were: Vichy mineral water (France), Mountain Valley Water (Arkansas), and Deer Park (New York).

Additionally, we sent samples of these selected waters to the Soil and Health Foundation, which, in conjunction with a private laboratory, tested them for coliform, pH, and trace minerals and metals.[12] Here's how the varieties compared:

Apollinaris

In a test of a random bottle of Apollinaris at a private laboratory, we found that it was salty—with a sodium content of 402.01 parts per million as compared with an average range in the United States of 1 to 198 parts per million. The test results also revealed arsenic levels of 0.300 milligram per liter—way above the EPA recommended limit of 0.05 milligram per liter. Also dangerously high, the cobalt level was measured at 0.114 milligram per liter. The EPA limit for cobalt in drinking water is zero. Selenium, too, exceeded EPA limits, measuring 0.185 milligram per liter as opposed to the suggested limit of 0.01 milligram per liter.

Perrier

An analysis of our bottle of Perrier showed that it contained a hefty amount of calcium—118.2 parts per million. It also offered traces of magnesium, potassium, sodium, barium, and iron.

Saratoga

Formerly called Saratoga Vichy, this water is highly mineralized —and tastes it. Our bottle contained a whopping 541.02 parts per million of sodium, with traces of calcium, magnesium, potassium, barium, and iron.

Canada Dry Club Soda

Surprisingly, this artificially carbonated water has won taste tests among gourmet celebrities.[13] It contains modest amounts of calcium and magnesium, a trace of potassium, 166.5 parts per million of sodium, and tiny bits of iron and zinc.

Deer Park 100 Percent Spring Water

Most people find Deer Park exceptionally clean and refreshing. This water probably could be dubbed the drink of presidents, since Benjamin Harrison, Grover Cleveland, James Garfield, and William Howard Taft all traveled to the Deer Park Spa to take the waters.

Our analysis found this water to be free of almost all minerals, containing small traces of calcium, magnesium, potassium, and sodium.

San Pellegrino Pure Mineral Table Water

Bottled since 1899, this water has long been a favorite among Europeans. When analyzed, our randomly purchased bottle was found to have very substantial amounts of calcium (139.7 mg/l), magnesium (50.15 mg/l), potassium (3.05 mg/l) and sodium (41.71 mg/l). It also contained small traces of barium, iron, and mercury.

Vichy Mineral Water

Julius Caesar built a health resort at Vichy more than 2,000 years ago. Eventually destroyed by invading barbarian hordes, it was later rebuilt by Napoleon II. The water gushes from an aragonite rock which has been set inside a plate glass frame. The water then splashes down to marble shelves, where it is caught in stone cups and served to guests.[14]

Our analysis, and our tasters, found Vichy to be extremely salty. In fact, our sample contained an astounding 790.7 parts per million of sodium per liter. Vichy also had high amounts of calcium (92.85 ppm), magnesium (11.90 ppm), and potassium (7.28 ppm), and traces of arsenic, barium, chromium, copper, iron, manganese, mercury, selenium, and zinc.

Mountain Valley Water

Mountain Valley is probably the most famous water our country has to offer. It was mentioned in the diaries of Hernando DeSoto all the way back in 1541—not by name, of course—when he told fellow Spaniards that the Arkansas Indians peacefully shared these recuperative waters. It was the first bottled water to be franchised in this country —uniting strange bedfellows like William Randolph Hearst, the crusading publisher, and Richard Canfield—the man Hearst was crusading against—in ownership of the New York City franchise.

Dwight David Eisenhower was ordered by his doctor to drink Mountain Valley water, and Kate Smith and her entire entourage have been known to quaff a quart or two.[15]

If our rather casual test proved anything at all, it is that European mineral waters taste too strong to please American palates. Of course, as they increase in popularity, we may acquire a taste for them.

The criteria you should use to choose bottled water for your family are very simple: the taste should be agreeable to your palate and the price agreeable to your pocketbook. If you want to drink only untreated bottled water, read all labels carefully. Natural mineral waters from deep wells and spring waters are clearly labeled.

If the label does not give a clear indication of whether the water has received any treatment, look for the address of the bottling plant. If the water has been bottled in a city, you can presume it has been treated. For example, the Hinckley & Schmitt Company is in the middle of Chicago and bottles supertreated Lake Michigan water. While this product is pure, it was treated to make it so.

Because Murphy's Law ("Anything that can go wrong *will* go wrong") affects the bottled water industry, just like any other, the best choice for bottled water is one that hasn't been treated at all. If you can find that the source of the product is a spring or well and that the water has emerged from the ground in a pure state, buy that brand if you can afford it. Look for labels that call the product pure spring water.

Bottled water comes in all sizes, from the 6-ounce Perrier, the quart bottle of Canada Dry, to a 5-gallon carboy of your local brand. If you use the smaller bottles—from pints to quarts and gallons—keep them stashed in your refrigerator. Water tastes fresher and more satisfying if it's cold.

Many families prefer to buy water in 5-gallon carboys. Some bottled water companies offer carboy service where you can arrange to have these big bottles delivered to your home, and replaced on a regular basis. Often a dealer will supply a metal stand free. It holds the jug so that it can be tipped for pouring. Others supply a dispenser—the old-fashioned kind that holds the bottle upside down so that it glugs whenever anyone taps a drink. Water from these dispensers can be refrigerated in a pitcher, or better yet, in a covered jar.

The ultimate in water dispensers is a unit that chills the water. With the push of a button, clear, cold, and satisfying water is there for the taking. Prices for these units vary according to location. Generally, they are sold or rented for use in stores and offices, but if you can pay the price you can have one installed in your home.

Those who live deep in the country may have trouble finding a dealer who will pick up and deliver on a regular basis. However, with a phone call to the local water company you usually can arrange a schedule so that you can pick up and return a carboy on your regular trips to town.

Some Fast Tricks

There are techniques you can use to clean your water of certain substances. Generally, the fast tricks listed below are really interim solutions to help you until you get your private water works in full operation. These will be especially useful to people who do not want to buy any bottled water.

Fast Trick #1

First thing in the morning, let your water run for two or three minutes at full force. By doing this, you will clear the water that has remained in your household plumbing overnight. Flushing out the system, you will reduce the amounts of lead, cadmium, and cobalt that may be in your water in substantial doses. While some traces of these metals may remain, chances are you have avoided giving yourself and your family a good shot of lead. Although many of us are groggy in the morning, once this procedure becomes part of the daily ritual it will become automatic.

Fast Trick #2

To get rid of bacteria in small amounts of water, you can boil it for twenty minutes. Or you can use a product called USP-grade resublimed iodine crystals. Buy 4 to 8 grams. It comes in a clear glass bottle with a hard plastic cap, and will cost $2 or $3. Since iodine in large quantities is a poison, you may have to ask your doctor for a prescription before your pharmacist will sell it to you.

Put a liter of water in a covered, closed container. Then, fill the iodine bottle with water, cap it, and shake it up and down for about a minute. Some of the iodine crystals will dissolve into the water,

Table 6-1: How Much Iodine to Add to 1 Liter of Water

Temperature (°F)	Volume (cc[a])	Concentration (ppm[b])	Capfuls[c]
37	20.0	200	8
68	13.0	300	5
77	12.5	320	5
104	10.0	400	4

Source: Fredrick H. Kahn, Barbara R. Visscher, "A Simple, Safe Method of Water Purification for Backpackers," *Western Journal of Medicine*, May, 1975, pp. 450–453.

a. cc = cubic centimeter.

b. ppm = parts per million.

c. Assuming a capful from a 1-oz. glass bottle holds 2 1/2 cc.

Table 6-2: Iodine Concentration vs. Time Required at 77°F

Concentration (ppm)	Time Required (min)
2	30
3	22
4	17
5	12
6	10

Source: Fredrick H. Kahn, Barbara R. Visscher, "A Simple, Safe Method of Water Purification for Backpackers," *Western Journal of Medicine*, May, 1975, pp. 450–453.

making an iodine solution. Allow the remaining crystals to settle at the bottom of the jar, and pour off the iodine solution into the jar's little cap. Carefully pour this capful into your drinking water container. Then let the treated water rest until the iodine disinfects it. The longer you can allow the water to rest, the less iodine solution you need to add. Also, the temperature affects how much iodine you will have to use. Iodine will destroy amoebae, their cysts, bacteria and their spores, algae and enteroviruses at 77°F in fifteen minutes. At near freezing, disinfection will require one and a half to two hours at the same concentration.

The procedure can be repeated about a thousand times with the same jar of iodine crystals, making the procedure very economical. However, it is time consuming and treats only small quantities of water. This procedure is an excellent precaution for campers and other travelers who are unsure of the local water supply.[16] However, iodine in large amounts is poisonous. For that reason, this method should be used only as a short-term measure, and not at all during pregnancy.

Fast Trick ♯3

Try the "Coffee Filter Method" using carbon. Granulated activated carbon (GAC) is very porous, and has an enormous surface area. Molecules of impurities will get trapped in it. Activated carbon is effective in removing pesticides, chloroform, and other organic contaminants. It also removes bad taste and odor. Filtration with GAC will reduce some heavy metals, but it will also reduce good elements like zinc and chromium. It won't remove nitrates or the essential minerals calcium and magnesium.

If you know you have a problem with organic chemicals in your water, put together one of these simple filters right away.

The directions for this filter were developed by Dr. Robert Harris of the President's Council on Environmental Quality—the same fellow who correlated the high level of organic chemicals of Mississippi River drinking water with increased cancer rates.

The materials you need include a large funnel, coffee filter paper, a glass collecting jar (a quart or larger) with a lid, and granular activated carbon. The carbon can be bought from Walnut Acres, Penns Creek, Pennsylvania. At this writing, the price of a pound was $2 (plus postage), and will buy about a three-to-six-month supply.

First, wash the carbon: put it into a jar, fill it with water, allow the carbon to settle to the bottom. Pour off the water, and continue the procedure until the water is clear. Now place the filter paper in the funnel along with enough carbon to fill the funnel about one-quarter full. Put the funnel into the collecting jar and begin very *slowly* pouring water into the funnel. The water must remain in contact with the carbon for the contaminants to be removed, so keep the stream down to a steady trickle. You can place a clean filter paper over the carbon so it will not be disturbed during the pouring. When the jar is full, bring the filtered water to a boil. It is important to prevent too much water from evaporating, so boil it in a closed tea kettle, and barely simmer it for fifteen to twenty minutes. Evaporation will reduce the amount of water, and result in a concentration of salts. Store the water in the refrigerator until you're ready to use it.

Fast Trick #4

Beat the chlorine and chlorinated organics out of your water. Put a small amount of water in your kitchen blender, and whirl it up for about fifteen minutes. Because gas is volatile, it will escape into the air.

Fast Trick #5

To counter chlorine's damaging effects on red blood cells, try vitamin C. Add a scant pinch of vitamin C powder or a piece of vitamin C tablet to a glass of chlorinated water immediately before drinking. Taste and odor will disappear.

This method works because vitamin C is an acid (ascorbic acid), while chlorine is a base. The acid combines with the base to form a salt. Therefore the chlorine is made innocuous.

Fast Trick #6

You will need the following materials: standard glass bulb baster (8¼ in × ¾ in), block of wood, rubber band, 1-quart glass jar, plug of cotton, activated carbon granules.

1. Place the granular activated carbon in glass jar. Fill with water, cover, and shake. Let the carbon settle to the bottom. If any fine black particles are present in the water above the settled carbon, pour off the water and repeat the procedure until the water is clear.

2. To construct the filter: drill a hole in the block of wood three-quarters inch or slightly larger to allow the glass bulb baster (without the bulb) to slide through. Place the rubber band on the glass baster just above the wood in order to prevent the column from sliding all the way through the wood. Set the block of wood on the mouth of the quart jar and suspend. Put a plug of cotton in the bottom of the baster and then fill half full with granular activated carbon.

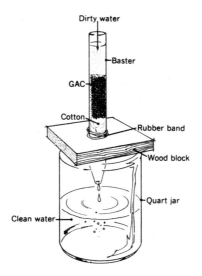

Figure 6-1: Fast Trick #6.

(Source: Birthright Denied: The Risks and Benefits of Breast-feeding, by Stephanie G. Harris and Joseph H. Highland [Washington, DC: Environmental Defense Fund, 1977], p. 52, by permission of the author and publisher.)

3. Pour tap water through the column.

4. When the jar is full of water, boil the filtered water by bringing to a boil, then barely simmering it for fifteen to twenty minutes in order to prevent too much evaporation and hence concentration of the salts.

5. Store in the refrigerator until used.

NOTES

1. "The Bubbly Fizz Biz," *Newsweek*, June 18, 1979, p. 69.

2. Arthur von Wiesenberger, *Oasis: The Complete Guide to Bottled Water Throughout the World* (Santa Barbara, CA: Capra Press, 1978), p. 17.

3. American Bottled Water Association, *Bottled Water Industry: Fact Sheet* (Los Angeles: American Bottled Water Association Headquarters, 1974), p. 2.

4. Ibid., p. 5.

5. von Wiesenberger, *Oasis*, p. 54.

6. "The Bubbly Fizz Biz," *Newsweek*, p. 69.

7. Department of Health, Education and Welfare, Food and Drug Administration, "Proposed Rule Making: Bottled Water, Proposed Quality Standard," *Federal Register*, January 8, 1973, p. 1019.

8. *Bottled Water Industry: Fact Sheet*, p. 2.

9. Department of Health, Education and Welfare, Food and Drug Administration, "Processing and Bottling of Bottled Drinking Water: Proposed Current Good Manufacturing Practice Regulations," *Federal Register*, November 26, 1973, p. 32,563.

10. *Bottled Water Industry: Fact Sheet*, pp. 2–3.

11. Ibid., p. 3.

12. Soil and Health Foundation, 33 East Minor Street, Emmaus, PA 18049.

13. "The Bubbly Fizz Biz," *Newsweek*, p. 69.

14. von Wiesenberger, *Oasis*, pp. 120–121.

15. Ibid., pp. 58–60.

16. Fredrick H. Kahn and Barbara R. Visscher, "A Simple, Safe Method of Water Purification for Backpackers," *Western Journal of Medicine*, May, 1975, pp. 450–453.

7

Equipment

If your concern about the quality of your drinking water is great enough for you to consider installing purification equipment, you will find many methods and choices of equipment open to you.

At first, the job of purifying water on your own may seem enormous, requiring the skills of a gifted chemist or engineer. But it's a job almost everyone can do. Remember, people have been purifying their drinking water since the dawn of civilization. The Egyptian Sanskrit, written about 2000 B.C., directed people to "heat foul water by boiling and exposing to sunlight and by dipping seven times into it a piece of hot copper, then to filter and cool in an earthen vessel." The Sanskrit claimed these directions came directly from "the god who is the incarnation of medical science."

Hippocrates, the father of medicine, wrote the first treatise on public hygiene, called *Air, Water and Places.* In it, he counseled public officials to consider the quality of water, saying ". . . for water contributes much to health." He recommended the use of a cloth bag to strain out impurities. It later became known as "Hippocrates' sleeve."[1]

In Egypt, infiltration channels were used to draw water from the lakes, purifying it of mud and leeches along the way. In Persia, water was boiled and stored in silver flagons for the Persian kings. Others recommended filtering water through sand, and even through wool. A popular chemical treatment that persisted for many centuries was the addition of wine to water. It is possible that, by adding strongly colored wine, drop by drop, the water's hardness could be determined.[2]

134

You do not have to treat the entire stream of water used in your home each day. You have to treat only the water you and your family drink and use in food preparation. Generally, that amounts to just 5 percent of the total.

The EPA estimates the average person drinks about 2 liters of water a day, and that estimate is considered a bit high.[3] Therefore, you likely have to purify no more than half a gallon of water for each person in the family each day.

There are many ways to treat water, and we will discuss just about all of them. But whichever method you choose, you will want to keep some criteria in mind. The unit you build or buy should perform all or most of the following tasks:

- leave the water free of harmful bacteria, viruses, amoebic cysts, and other parastic organisms
- not breed and release harmful bacteria into the water
- remove off-colors, odors, and tastes
- remove sediment
- remove microscopic asbestos fibers
- remove chlorine
- remove chloroform and other THMs
- remove organic chemical pollutants
- remove harmful heavy metals
- moderate or remove iron, manganese, or hydrogen sulfide
- provide sufficient amounts of water to supply a family's needs in a reasonable time
- not consume large amounts of energy
- indicate when maintenance is required or when the useful life of the treatment unit is over[4]

Among the devices on the market, you can choose from a variety of filters, distillers, deionization units (like water softeners), and reverse osmosis membranes. In addition, you can consider disinfecting water with ozone, ultraviolet light, or with chlorine. Some devices are limited in their purification capabilities, while others do a good general job of eliminating the most common problems. Let's look at each.

Water Filters

Simple sand filters are much the same as those used by the ancient

Egyptians. They are simply a column of sand or other porous matter that strains out particles from water. This kind of filter can remove clay, silt, colloids, and microorganisms, ranging in size from 1 millimicron to 50,000 millimicrons. Each bacterium and virus can be strained out of water, if the pores in the filter are small enough.[5]

The most widely used filters are sand filters in which 18 to 30 inches of sand are used, supported by a gravel layer about 6 to 12 inches thick. Slow sand filters are commonly used for pond water treatment because the cost is low and the effectiveness high, if they are properly cared for. Good maintenance consists of frequent backwashing to remove the accumulated dirt and scum on the top inch or 2 of sand. A slow sand filter can handle about 2 or 2½ gallons of water per square foot of sand surface, per minute.

Some of the substances you can use instead of sand include: crushed anthracite, alluvial anthracite, diatomaceous earth, gravel, and activated carbon.[6] These media can be used individually or together in a wide variety of arrangements. Filters that used several kinds of material are called multimedia filters (generally they are made with the largest particles, like gravel, at the bottom and layered upward with increasingly smaller material to the top, which is made of the finest material, silica sand). With this arrangement, the large particles support the finer ones, keeping them from slipping down the drain.

However, some multimedia filters are inverted, with larger particles on top. This arrangement allows the larger coarse material to be filtered out of the water before it comes in contact with the smaller media particles. The smaller particles are spared from being clogged with coarse matter and thus are able to remove the fine turbidity for longer periods between backwashing. Use of inverted multimedia filters provides "in-depth" filtration, where the filtration process takes place throughout the entire depth of the column. In contrast to conventional multimedia filters where the top few inches have to filter out the entire contents of the water, an inverted filter will last about five times longer before it clogs and needs backwashing. It also can handle a much faster flow of water because the thick dirt layer does not build up on the large gravel top and therefore it does not restrict the flow.

While water moves through filters by gravity in most designs, some filters depend on pump pressure to either push or pull the water through it faster.

At the bottom of the filter is a drain which removes the cleansed water. It is simply a pipe punched with minute pores to allow it the passage of water.

Filter maintenance is required when the accumulated scum, called filter cake, begins to slow down the water flow. When this "filter cake" is about 1 or 2 inches thick, it has to be removed along with the top inch of sand (or whatever medium is on the top layer). When about 6 inches of material have been removed, the filter should be topped off with fresh sand.

Filters are so simple they can be easily constructed at home.

Constructing a Simple Filter

This filter is designed to remove both harmful bacteria and suspended material. It was tested in southern Iran, and found to be simple and effective. In this country, its best use is probably as an initial treatment for pond water that will be used for drinking. This cleansing will make the water more suitable for further treatment in a distiller, GAC filter, reverse osmosis unit, or deionizing unit, and make maintenance of these units less frequent.

To make this filter, you will need a 55-gallon drum with a lid. First attach a drain faucet at the side near the bottom. Assemble your filter media and wash all of it to remove dirt and dust. Now put in 6 to 8 inches of gravel of sizes varying between 0.5 and 2 inches. Place 6 to 7 inches of activated carbon on top of the gravel. This will help support the sand on the next layer and keep it out of the drain. Between 27 and 36 inches of sand are needed.[7] The sand used should consist of hard durable grains free of clay, loam, dirt, or organic matter. It should have an effective size of 0.20 to 0.40 millimeter. (The effective size is the diameter of the grains in millimeters, which make up 90 percent or more, by weight, of the sand.) This layer should be covered with 4 to 6 inches of gravel. The net result is that about two-thirds to three-fourths of the drum should be filled (see Figure 7-1).

Turbid water is poured into the top of the drum, and filtered water comes out at the bottom.

Such a filter will become more efficient at removing turbidity as it "matures." The mat of turbid materials that builds up on the top layer promotes better filtration naturally. The thick layer of sand will

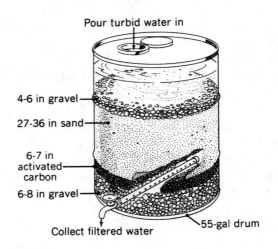

Figure 7-1: Simple Multimedia Filter for Home Construction.
(Source: Patricia M. Nesbitt, Environmental Consultant, Strasburg, VA.)

help reduce the bacterial levels considerably. Maturation comes in 15 to 17 weeks, and efficiency is maintained for another 6 to 100 weeks. Thereafter the filter becomes saturated, the rate of flow slows down, and turbid matter bleeds through. At this point, the filter content can be dumped, dried under the sun, dusted, and well rinsed. The carbon should be replaced, but the sand and gravel can be reused.

Filters That Fight Corrosion

In previous chapters, we discussed the problems of acid water. It is known that acid water corrodes water lines and plumbing, often adding heavy metals to our drinking water, creating potential health problems while at the same time reducing the life expectancy of household plumbing.

To neutralize acid water, you can build a filter that consists of limestone or marble chips. As water passes through the filter bed, the acid is neutralized by the limestone, forming bicarbonates.[8]

These filters are easy to use and require little care. Weekly backwashing and an annual replacement of dissolved chips is all that's required.

Another method would be to attach a pump to the well pump which pumps a solution of potassium carbonate into the water every time the pump comes on. This will neutralize the acidity, and will add potassium, which is probably in short supply, if anything, in most of our diets. There is no problem with bacterial growth in the solution of potassium carbonate, because the pH of the solution is so high.

Filters That Remove Chloroform and Other Organic Chemicals

Activated carbon has been found to be extremely effective in removing chloroform, chlorine, some pesticides and other organic chemicals, as well as bad taste and odor.[9] Fast Tricks #3 and #6 told you how to make small filters with granular activated carbon as a stop-gap measure. It is possible to construct a full-scale GAC water filter on your own, or to purchase a commercial unit.

Activated carbon filters (mistakenly called charcoal filters by some) are the least expensive and most common of the dozens of new water treatment gadgets on the market. The most popular units—and the least effective—are the small filters that attach to the end of the faucet and filter water as it leaves the tap.[10] Larger units can and do perform effectively.

Large or small, these filters all contain activated carbon, a substance used in medicine since at least 1550 B.C. Later, sailors learned that drinking water stored in charred barrels stayed fresh during long journeys. In fact, charred barrels often improved the water's taste. But carbon's use in water treatment didn't become widespread until the twentieth century, when it was added to sand filters similar to the one described previously. Today activated carbon is not only used in water treatment, but also in air pollution control devices, purification of chemicals and gasses, dry cleaning, cigarette filters, and gas masks. It is also key in the treatment of water used by the beverage and food industries, and in the purification of water used in kidney dialysis units. Its entry into the home is not new by any means. Activated carbon is the familiar black material you put in your fishtank filter, or use to clean the water for your steam iron.

Activated carbon is a form of charcoal that has been treated by high temperatures and steam in the absence of oxygen. This process produces small granules that are extremely porous with a high surface area on which organic molecules can cling and are "adsorbed." The surface area is huge—with 1 pound of granular activated carbon providing a surface area equivalent to 125 acres. *Adsorption* is a scientific term referring to the process of adhesion or sticking of dissolved substances to the surface of the carbon particles. An activated carbon filter does not actually filter the organics out of the water—or soak them up like a sponge—rather it holds them on its surface.

Granular activated carbon (GAC) provides home treatment with great efficiency of removal. When GAC is used in a tall column, most of the unwanted organic pollutants are adsorbed in the top layers, and as the water passes downward it progressively contacts cleaner and cleaner carbon. This in-depth filtering "polishes" the water as it passes out of the column and provides a wide margin of safety.

GAC has the capacity to remove or reduce many organic chemicals, including pesticides, industrial chemicals, and most halogenated organic compounds like PCBs and PBBs (polychlorinated and polybrominated biphenyls, respectively). Chlorine and by-products of chlorination such as chloroform are also removed. While nobody knows for sure if GAC will remove all organic chemicals from water, it is the best treatment we have today.

Carbon also will reduce the concentration of heavy metals like lead and cadmium, but at the same time it also reduces the good guys like zinc and chromium. It will not remove fluoride, nitrates, or other salts, nor will it remove asbestos fibers. In fact, if water is loaded with minute debris, it is best to use a sediment prefilter before putting it through carbon, or the carbon will become clogged and quickly lose its effectiveness. One other limitation is that the adsorption property of carbon filters changes with water temperature. This limitation probably has the most impact on the little end-of-faucet filters that receive both hot and cold water.

While a wide variety of filters are made with GAC, not all will perform equally. Effectiveness varies with the kind of carbon used in the filter, the filter design, the way it is used and maintained, its age, and the condition of the water to be treated.

Carbon cleans best when water slowly passes through a large mass of it. The best systems employ a long column filled with carbon through which water drips at a rate of about 2 drops per second—or just enough to keep the carbon wet. The tall column and slow rate allow water to make intimate contact with the carbon, thereby allowing for maximum chemical removal.

A filter's effectiveness in removing organics also depends on how long the carbon has been used, how much water it has treated. GAC has a limited capacity to adsorb organic molecules because after some use the carbon surface becomes saturated. If too much water is passed over the carbon, the organics already on the carbon can break off, contaminating the water. And nobody can predict exactly when this action is likely to happen. However, this release appears to be gradual and does not produce "slugs" of organic waste in the drinking water.

But the major problem with GAC filters is that bacteria and other organisms grow on the carbon surface. Generally, these bacteria are not the kind that cause typhoid or cholera, but in large doses they can cause intestinal upsets. And they *do* come in large doses. Alarmingly high levels have been found in the first water to come through the filter after an overnight rest. The quiet water and the damp carbon create good conditions for bacterial growth.

Some manufacturers add silver compounds to their filters to kill bacteria. However, health-conscious consumers should look for filters without silver, because preliminary testing has shown that the silver is not a dependable bacteriostat. The consumer can be lulled into a false sense of safety if he believes the silver is doing a good clean-up job, when, on occasion, it may not be doing anything of the sort. Additionally, silver can be harmful to health, and with some filters small amounts of silver are leached into the treated water.[11]

Because of the threat of bacterial and chemical recontamination, GAC filters require scrupulous maintenance. Carbon should be replaced every three weeks, or after treating 20 gallons of water. In the interim, water should be monitored closely for signs of filter failure. Here are some signs that carbon replacement is long overdue:

• A change in taste means the carbon bed is saturated, and harmful chemicals may be passing through it.

• The water pressure is noticeably reduced, resulting in slower than normal output. Reduced pressure means the filter may be clogged, possibly with bacteria, and is not adsorbing chemicals very well. If you *must* drink this water, boil it first. Otherwise replace the carbon.

• When small particles appear in the water, it means the filter is clogged. Replace the carbon.[12]

Remember that, if the carbon isn't properly maintained, you are creating the same sort of problem you've been trying to lick—impure water. Researchers at the Baylor College of Medicine ran tests on local tap water that was held in four sink-top carbon filters. Overnight the bacterial count rose from several hundred per 100 milliliters of water to seven *million*. [13]

Which Is Best?

Testing of home filters using GAC has been very limited. Some of the advertising claims made by a few filter manufacturers are outrageously misleading. In fact, in 1978 the Environmental Defense Fund petitioned the Federal Trade Commission to investigate these claims. Dr. Ervin Bellack, a water chemist at the EPA, has said, "I've seen some home filter advertisements that say you can run sewage through them and get water pure enough to drink. What I'm afraid of is that someone will take these claims seriously."[14]

In 1979, the EPA began preliminary testing of a few GAC filtration units. They chose to examine the capabilities of several types ranging from a simple pour-through unit to a double-filtration system. The tests were made to see just how effective GAC can be in removing contaminants. While many of these units never promised to do more than improve the taste and odor of drinking water, the EPA nevertheless subjected them to rigorous testing to see how well each type performed.

In these preliminary tests, seven units were chosen. They included: the Instapure Model F1-C (faucet bypass); Mini Aqua Filter (faucet, no bypass); Sears Taste and Odor Filter (stationary); H₂OK Portable Drinking Water Treatment Unit (pour-through); System 1 Water Processor, Model SY 1-34 (line bypass, silver); Aqualux Water Processor, Model HB (line bypass, silver); and Everpure, Model QC4-THM (line bypass).[15]

The models listed as containing silver simply have silver-impregnated activated carbon to inhibit bacterial growth. Models with faucet bypass allow water to pass through the tap directly, bypassing the filter.

The EPA tested these units in several ways. It tested them with New Orleans tap water, and again with tap water spiked with an additional 200 parts per billion chloroform, to simulate the condition of some drinking waters.[16] The units were tested to see how well they removed the organic chemicals in the water.

The filters were also judged on the way they removed a substance called nonpurgable total organic carbon (NPTOC), which represents the bulk of organic compounds in drinking water, and is believed to be mostly natural materials that have a high molecular weight.[17]

In addition, they were tested to determine how well they removed bacteria, and a substance called endotoxin. Endotoxins are given off by some bacteria as they die, and can remain in drinking water. It is known that endotoxin can cause fever when injected into a person's blood stream, but the health effects of endotoxin in water are not yet known.[18]

In removing THMs, the faucet unit with no bypass removed only 6 percent; the faucet unit with a bypass removed about 25 percent; the three line bypass units removed from 43 to 93 percent of all THMs; the stationary unit removed 46 percent; and the portable pour-through unit removed 19 percent[19] (see Figure 7-2).

Tested for bacteria, the water that came out of these units had more bacteria than the water that went in. However, there were rapid shifts and reversals in the test results. The water was also tested for

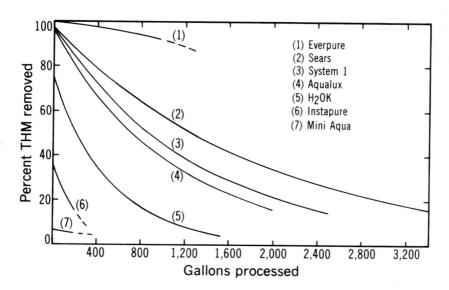

Figure 7-2: Summary Performance of Carbon Units for THM Removal.
(Source: Development of Basic Data and Knowledge Regarding Organic Removal Capabilities of Commercially Available Home Water Treatment Units Utilizing Activated Carbon: Preliminary Report—Phase 1 *[New Orleans: Gulf South Research Institute, 1979], by permission of the publisher.)*

Table 7-1: Trihalomethane (THM) and Nonpurgable Total Organic Carbon (NPTOC) Removals EPA/GSRI Contract

Type[a] I.D.	Name	Run I.D.[b]	Test Life (gal)	Est. Ave. Cont. Time (sec)	THM Ave. Infl. Conc. (ppb)	THM Net Overall Removal (%)	NPTOC Ave. Infl. Conc. (ppb)	NPTOC Net Overall Removal (%)
(A)	Instapure F1-C	1 2	200	1.6c 0.5d	213 260	24 25	2,193 2,642	11 11
(B)	Mini-Aqua Filtere	1	200f	0.9	213	6	2,193	2
(C)	Aqualux Water Processor Model HB	1 2	2,000	23 23	223 348	45 45	2,252 2,664	28 28
(C)	Everpure Model QC4-THMg	2	1,000	43	345	93	2,573	41
(C)	System 1 Water Processor	1 2	2,500	46 46	232 347	43 43	2,244 2,630	20 20
(D)	Sears Taste and Odor Filter	1 2	3,420	4.5 4.5	229 368	46 46	2,164 2,682	12 12
(E)	H2OK	1 2	2,000	15 15	234 354	19 19	2,522 2,595	10 10

(The table headers, spanning: "Information on Units Tested", "Trihalomethane Removals (THM)", "Nonpurgable Total Organic Carbon Removals (NPTOC)")

Source: *Development of Basic Data and Knowledge Regarding Organic Removal Capabilities of Commercially Available Home Water Treatment Units Utilizing Activated Carbon: Preliminary Report—Phase 1* (New Orleans: Gulf South Research Institute, 1979).

a. Code for type of units: (A) faucet/bypass; (B) faucet/no bypass; (C) line bypass; (D) stationary; (E) portable pour-through.

b. Code for run I.D.: (1) tap water (no spike); and (2) tap water spiked with 200 ppb chloroform.

c. Based on using manufacturer's recommended flow rate of 0.5 gal/min.

d. Based on using experimentally determined flow rate of 1.5 gal/min.

e. Tests were discontinued after the tap water (no spike) tests because removals of THMs and NPTOC were negligible.

f. Since the manufacturer claimed an indefinite life span for this faucet unit, 200 gal were taken for test purposes, as being compatible with other faucet units.

g. Not tested in tap water (no spike) runs but added for the spiked tap water runs.

coliform bacteria, and none was found. Researchers also found that none of the silver impregnated units consistently demonstrated bacteriostatic properties.[20]

Generally, the tests showed that stationary and line bypass units can reduce trihalomethane concentrations. The Everpure unit, which reduced the concentration by 93 percent, was designed specifically for this purpose. It employs two filter cartridges, one of granular carbon and the other of powdered carbon.[21] It is, however, expensive to buy and maintain.

The small faucet filters with only a small amount of carbon were shown to have a limited capacity for THMs. Generally, the contact time of water with carbon was also very brief, ranging from one to fifteen seconds. The tap water simply rushes through too fast for the GAC to do its work. One unit, the Mini Aqua, showed such inefficiency in removing THMs that it was taken out of the test altogether.

While these test results are very preliminary, they reveal that the basic premise is sound. To have an effective GAC filter, water must pass through sufficient quantities of carbon, and it must pass through it slowly. If these two prerequisites are met, the filter should work well.

Make Your Own Carbon Filter

This filter has been designed by the Environmental Protection Agency, and was tested on water from the Ohio River in Cincinnati —one of the foulest sources of drinking water known to mankind. It produced high-quality drinking water for three weeks, before the carbon needed changing.

As you can see from Figure 7-3, a saddle valve is used to tap the water at a rate of only 1 gallon a day, to ensure a good long contact time between the water and the carbon. The filter should operate continuously, twenty-four hours a day, at a rate of 1 gallon of filtered water a day.

The carbon filter should be changed every three weeks, or after approximately 20 gallons of filtered water. To be absolutely safe, the filtered water should be boiled gently for fifteen or twenty minutes in a covered pan, and stored in the refrigerator.

When changing the carbon, disinfect the rest of the filter by soaking it in a 5-percent solution of laundry bleach.

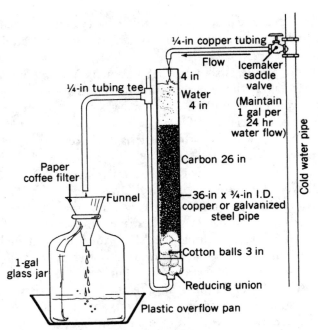

Figure 7-3: Self-Leveling Activated Carbon Column.
(Source: Walnut Acres, Penns Creek, PA 17862.)

This homemade unit can be easily installed in a cabinet beneath the kitchen sink.

For materials you will need ¼-inch copper tubing, ¼-inch tubing tee, 36-inch by ¾-inch inside diameter copper or galvanized steel pipe, an ice maker saddle valve, reducing union for ¾-inch pipe to ¼-inch tubing, funnel, paper coffee filter, 1-gallon glass bottle, plastic pan, and cotton balls.

1. Construct the filter as illustrated. A shorter column can be used if there are space limitations.

2. Disinfect the empty column with 5 percent solution of laundry bleach by filling the column and letting it stand for a couple of minutes.

3. Rinse the column thoroughly.

4. Add the cotton plug.

5. Fill the column with water and add previously wetted (two hours), washed carbon to the top to a depth of 26 inches.

6. Maintain the water level above the carbon by placing the ¼-inch tubing tee above the surface of the carbon as illustrated.

7. Operate the carbon column continuously, using the saddle valve to adjust the flow rate.

For absolute safety, boil the water and store it in the refrigerator.

Remember to change the carbon every three weeks or after filtering every 20 gallons.

Distillation

Distillation is a time-honored method of treating water. Water is heated until it turns to steam, and then the steam is condensed into water. In theory, all the debris, bacteria, minerals, and other contaminants are left behind as the water turns to steam, and what is condensed, therefore, is very pure water. Generally, the theory holds true but it isn't altogether accurate. Some contaminants, namely, chloroform and other organic chemicals, have a lower boiling point than water. They can vaporize right along with the water, recondense, and wind up in the finished product. However, there are types of distillers, called fractional distillation units, that prevent this from happening. If you decide to buy one, be sure it is a fractional distiller. Many are not.

The basic components of a distiller are the heater, the boiling chamber, the vaporizing column, and the condensing chamber. A collection bottle is also required.

Within the boiling chamber, incoming tap water is heated almost to boiling. All impurities which do not boil below the boiling point of water ($100°C$ or $212°F$) are left behind in the chamber. Other impurities with a lower boiling point than water (like chloroform) will vaporize quickly, and are removed from the unit before they condense. In this way, the vapors of chemical impurities are left behind. Other stills have a gas trap at the top of a shorter column, where vapors can be vented.

After the steam rises in the column, it moves into the condensing chamber where it is cooled and returns to its liquid state. It is important that no leaks develop in the tubing, because leaks would allow the passage of contaminants into the purified water. It is also important that the tubing is made of something that will not contaminate the water. Glass and stainless steel are best.

The steam moves down the condensing tube, which is surrounded by a cooling jacket. Either cool air or cool water is circulated in the

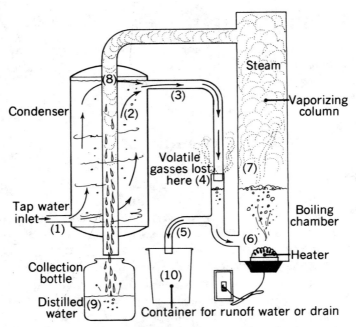

Figure 7-4: The Fractional Distillation Process.

(Source: Joseph P. Senft, Ph.D., Research Scientist, Organic Gardening and Farming [Research Center, Rodale Farm, Maxatawny, PA].)

How it works:

(1) Water enters the cooling jacket via the tap water inlet.

(2) By regulating the inflow rate, water leaving the cooling jacket is heated to nearly boiling by the heat from the steam condensing within the condensing chamber.

(3) As the water flows into the boiling chamber, it is vented (small pipe flows into big pipe).

(4) Gasses with a boiling point lower than water will be lost to the air at this point.

(5) Any substance with a boiling point greater than water is carried out in the drain.

(6) The water flows to collect around the heater.

(7) 10 to 20 percent is vaporized. The remainder flows out the drain.

(8) Water in the cooling jacket acts on the steam coming from the vaporizing column, condensing it into drops of distilled water.

(9) Drops of distilled water are collected in a container at the bottom of the condenser.

(10) The runoff water, which is hot, may be collected and used for dishwashing or laundry, or discarded by positioning the outlet above a household drain. Note: Solids such as salts (calcium carbonate) collect on the walls of the vaporization chamber and must be cleaned out periodically. Cleaning frequency depends on the total "solids" content of the water.

cooling jacket. As the steam cools, it forms tiny droplets along the inside of the condensing tube. The droplets combine, and flow into a collecting chamber. Now the water is distilled and purified.

The same procedure is used to distill crude oil into gasoline, and corn mash into liquor.

Once the water is distilled, it should be stored in the cleanest container possible, and kept in the refrigerator. Old hands at distillation say the real trick is keeping good-quality water, not making it. The best containers are of glass and stainless steel. Do not use a plastic jug. Not all of them are inert—that is, they can contaminate your water during storage.

Start saving gallon glass jugs from cider, vinegar, or wine. Wash them with warm, soapy water, and rinse well. Then rinse them again with distilled water and white vinegar. Finally, rinse them one more time with distilled water. Once filled, keep the jugs in the refrigerator or at least a cool spot in the kitchen.

Distillation may not be the process for you. For one thing, some distillers are very hard to keep clean. When tap water is boiled, and the steam passes on to the condensation chamber, the minerals, chemicals, and other contaminants are left behind in the boiling chamber. Eventually, they build up to form a scale that can interfere with the unit's efficiency and even with the quality of the distilled water. For that reason, the units need frequent cleaning. One of the first problems you may encounter is getting to the boiling chamber. Some units make access very difficult. Others have a shape or size that makes it difficult to get into. Be sure to consider the design of any unit you purchase with this cleaning procedure in mind. Some distillers can be cleaned only by washing with a strong acid, like nitric acid. Others have automatic flushing systems that purge the residue from the chamber frequently. A third type has a boiling chamber that can be cleaned with a wire brush. Be sure the distiller you buy is one you will be able to maintain.

Secondly, stills use a lot of water. Generally, stills use 5 gallons of tap water to produce 1 gallon of distilled water because the water is used as a coolant as well as the source for the distilled water. Quantities and ratios vary with individual units. If you do not have a plentiful and steady supply of household water, distillation probably is not the purification technique for you. Moreover, if your water is hard, your distiller will clog quickly. However, the waste water from a distiller can be used in your garden, which will benefit from the concentrated minerals. Or, the hot water can be diverted to a washing machine for reuse. And the heat and humidity produced by a still can make a room very comfortable in the wintertime.

Thirdly, a distiller has to operate at the proper temperature to drive off chloroform. The heat sources for the boiler can be solar, gas, electric, or kerosene. Usually, a small heating element is contained in the larger distillation units. Some small stills, however, are designed to work on top of a kitchen stove. But some stove-top models have been known to burn out because the heat is not regulated carefully.

It's also very important to estimate the energy demand of a still. As the prices of electricity and natural gas continue to rise, the cost of running a still may become too expensive for the average family. When you begin to shop for stills, ask how long the process takes and multiply that by the cost of electricity in your area (per kilowatt hour). Many still owners argue that the heat requirement is really minimal. Once the water has been heated to the boiling point, very little extra heat is required to keep it there. Distillation units that regulate the heat flow —providing lots of heat initially and tapering off—are safe and energy efficient. Units that don't lower the temperature automatically have been known to burn out.

But aside from the amount of water and energy required by a still, the greatest drawback may be the taste of the water it produces. It doesn't have any.

This lack of flavor may take awhile to get used to, but most people adjust. The reason the water has no taste is that all the minerals have been removed during distillation. For those who may be concerned about losing that supply of minerals, additives are available. One such substance is called Liquid Chelate, sold by Nector-D'Or, Box 1405, Provo, UT 84601.

A final word about distillers. They are expensive, ranging from about $150 to more than $500.

There are two major fractional water distiller companies in the United States:

New World Distiller Corporation
Box 476
Gravette, AR 72736

Pure Water, Inc.
3725 Touzalin Avenue
Lincoln, NE 68506

Deionization

Ion exchange is a common chemical reaction. One ion is substituted for another in solution.

Water softeners work on the principle of ion exchange, with sodium ions substituting for calcium and magnesium ions. Once the calcium and magnesium are removed, the water is no longer hard.

A water softener consists of a tank containing an ion-exchange material such as zeolite or beads of plastic resin. Some designs have a second tank containing a brine solution to regenerate the ion-exchange material. When the water passes through the zeolite or resin beads, "hard" calcium and magnesium ions are exchanged for "soft" sodium ions. In this manner the water is softened. Some water softeners use a kind of zeolite material that also will remove soluble iron.

For water containing

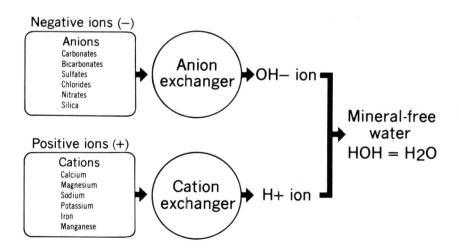

Figure 7-5: The Deionization Process.
(Source: Continental Water Conditioning Corporation.)

But water softeners are not the only kind of deionizing unit that can be used to treat water. There are others that use hydrogen, not sodium, to replace unwanted elements in water.

These deionization units have two ion-exchange materials—one to remove positively charged ions (called cations) and another to remove negatively charged ions (called anions).[22] The water passes through the first material and exchanges its cations for hydrogen ions. All cations—not just calcium and magnesium—are exchanged. At this point, the water is acid because of its high concentration of hydrogen ions (H+).

Although the cations have been exchanged, the negatively charged ions called anions still remain in the water until it is passed through the second material. Here the anions are exchanged with something called hydroxyl ions (OH−). When the hydrogen ions (positive charge) meet the hydroxyl ions (negative charge), they combine to form water: a new water that contains no minerals.

$$H+ \; + \; OH- \; \rightarrow \; HOH \; or \; H_2O$$

While this process may sound like hocus-pocus to those not familiar with it, it actually is a very old procedure first written about by Sir Francis Bacon.[23] Today, laboratories that require the very purest water use units that contain special exchange resins that are very effective and efficient. However, these resins have not been marketed for home units.

When it comes to cost, deionization is a very fair way to treat water. The harder or dirtier the water, the more it costs to clean it up. The reason is that ion-exchange resins have a limited capacity. As the exchange sites become filled with contaminant ions, the unit becomes less effective. The saturated resin must then be removed and regenerated.

The regeneration process varies with the type of resin used and the type of contaminant ion. Generally, regeneration involves the removal of the contaminant ion and its replacement with the H+ and OH− or Na+ and Cl− ion of that which was originally on the resin before it was put into use. Regeneration is accomplished by flooding the resins with extremely concentrated acid and base solutions of the original ion, and exchanging the ions again. The resulting concentrated waste then must be disposed of.

If you're wondering whether it will cost you a small fortune to treat your water, check over these guidelines for what water should contain:

- less than 5 parts per million of iron
- less than 5 parts per million of manganese
- less turbidity than 5 parts per million
- limited amounts of organics, since these can foul the unit
- chemical oxygen demand not more than 1 part per million
- maximum chloride of 0.2 part per million
- no hydrogen sulfide present in water to be deionized
- oil limited to 9 parts per million

These stringent requirements for untreated water may explain why deionization is not common on the household level.

However, deionization is much more reliable for mineral removal than distillation—depending on the individual units involved. Deionization will remove hardness, copper, fluoride, soluble iron, nitrates, silica and silicates, sodium, sulfate, total dissolved solids, arsenic, and selenium. However, the treated water is not sterile. More important, organic chemicals usually remain behind. One resin has been developed by the Rohm and Haas Company that does remove organic chemicals, and can be regenerated with steam or solvents. However, it is not on the consumer market at this writing.

Reverse Osmosis

Reverse osmosis (RO) units are relatively new, and most people still consider them a fairly exotic way to treat water. With reverse osmosis, water flows over the surface of a membrane that looks something like cellophane. Under pressure from the tap, some water is forced through the membrane, while the remainder which is now heavy with matter, remains to be drained away. It is a kind of filtration under pressure that can remove minute pollutants from the water.[24]

Distillation and reverse osmosis both involve removing water from its impurities. Other water treatments, like softening, filtration, and deionization, involve removing impurities from the water. Osmosis occurs when two solutions of different concentrations are separated by a semipermeable membrane. The water passes through the membrane

Semipermeable membrane   Semipermeable membrane

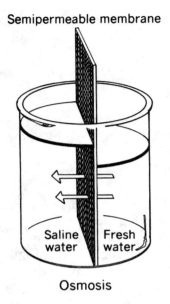
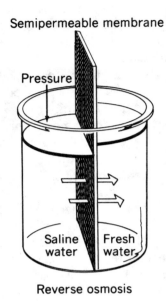

Osmosis   Reverse osmosis

Figure 7-6: Osmosis and Reverse Osmosis.

When osmosis occurs, water moves into the fresh water well, leaving salts behind. In reverse osmosis, water moves through a semipermeable membrane into a low concentration of nutrients (or salts) from a high concentration of salts when pressure is applied to the high concentration.

in the direction of the more concentrated solution. By osmosis, oxygen from our lungs enters our blood. By osmosis, plant roots take up water and nutrients. But reverse osmosis is exactly what the term implies— the reverse of osmosis. Instead of one solution being drawn into the more concentrated solution, the application of pressure causes the more concentrated solution to be forced through the semipermeable membrane. The membrane is permeable to water but not to its impurities. As a result the water passes through the membrane but the impurities do not. RO is similar to filtration, but will remove not only the matter in water, but, unlike a filter, also the *dissolved* matter in water.[25]

RO removes turbidity, particulate and colloidal matter, ionized and nonionized dissolved solids, bacteria, viruses, and pyrogen (fever-causing substances).[26] It will also remove most organic chemicals, or those with a molecular weight of 200 or more.[27] Therefore, it will remove aromatic hydrocarbons, most pesticides, and other complicated chemicals. But it will not remove simple compounds like chloroform and phenol.

Reverse osmosis is also an excellent way to remove asbestos from water.[28] Even water spiked with extraordinary amounts of asbestos—200 million fibers per liter, in one laboratory test—was absolutely free of asbestos after RO treatment. In fact, the Army Corps of Engineers used reverse osmosis to treat the asbestos-laden water in Duluth.

Reverse osmosis is a simple process, both in concept and practice. All you need is the reverse osmosis membrane and the normal water pressure present in the service line. No additional energy is required to operate the unit. This is one of the greatest advantages of treating water with an RO unit.

Several types of units are available. Some have a membrane that is supported on a special backing that can withstand the high pressure while it allows for the passage of the cleaner water. Other designs use membranes in the form of hollow fibers, about the thickness of a hair. These fibers need no backing because of their great tubular strength. In addition to clean water, the unit produces as a by-product waste water of dissolved minerals and other debris removed from the water. This waste water can easily be diverted to a drain.

These membranes have a useful life of one to three years, varying with the pressure, pH, temperature, and quality of the water being

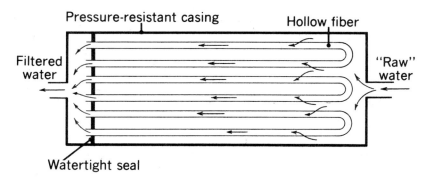

Figure 7-7: Cross Section of Hollow Fiber Device.
(Source: Lee Jaslow, Environmental Consultant, Baltimore, MD.)

Real units have thousands of fibers, each no more than 100 microns in diameter.

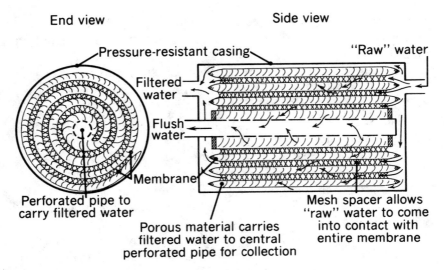

Figure 7-8: Cross Sections of Spiral Membrane Device.
(Source: Lee Jaslow, Environmental Consultant, Baltimore, MD.)

treated. If the water is oily, for example, the membrane will be fouled very quickly and need frequent maintenance. The pH of water passing through a membrane should be between 2.5 and 7, and the temperature should be below 100°F.[29]

Another limitation of these units is that they can develop problems from buildup and scaling. Some experts recommend pretreating very hard water with a water softener before passing it through a reverse osmosis unit.[30] However, most household water will not require pretreatment.

To provide the same sort of protection, some units are fitted with a prefilter which helps remove larger particles from incoming water. If this filter is used, it will have to be replaced every year or year-and-a-half.

Another option for RO units is to add a carbon filter, through which water passes after it leaves the membrane. This filtration assures good flavor, and odor-free water and gets rid of chloroform and phenols as well.

Because RO works slowly, the unit is usually run frequently in order to provide enough drinking and cooking water for the whole family. The processed water is stored until there is enough to meet demand.

RO units are designed primarily to remove dissolved solids. They can, for example, turn sea water into drinking water. In terms of ordinary tap water, RO will remove chlorides, fluoride, calcium, magnesium, manganese, nitrates, silica, silicates, sodium, sulfates, and copper. Detergents, organic matter, tannins (which can hide iron compounds), chlorinated compounds, and many toxic chemicals are also removed.

A reverse osmosis unit may cost from $300 to more than $500. However, the initial cost is offset by the fact that these units need no electricity or any other source of energy, as they operate with water pressure.

Probably the most popular RO unit for household use is made by the Culligan Company. Called the Aqua-Cleer, the unit fits in under a kitchen sink and runs the processed water to its own faucet. The unit also contains a fine sediment filter, a carbon adsorption filter, and even offers an option that heats or chills the water.

Aeration

Aeration is a simple, natural treatment of water. By spraying, bubbling, or thinly pouring water, it comes in intimate contact with air. This procedure reduces the concentration of gasses—like the rotten-egg-smelling hydrogen sulfide. It also causes volatile organic compounds like chloroform and methane to dissipate into the air. For such a simple procedure, it is quite effective.

Aeration can also be used to oxidize iron and manganese which will then precipitate out of the water as insoluble oxides, to remove odors produced by algae, and to brighten up the flat taste of cistern or distilled water.[31]

In general, all gasses and dissolved compounds which are easily volatilized will be removed through aeration. Many of the hydrocarbons dumped into rivers and streams by careless industries will volatilize with aeration. Chloroform and other simple chlorine compounds may also respond well to aeration depending upon temperature and initial concentration.

However, whatever is present in the air can be added to water during aeration. Oxygen is an obvious contributor, and an increase in oxygen will improve the taste and odor of aerated water. But any airborne contaminant can also be incorporated into water. For this reason, it is wise to be careful about where you aerate water. A closed room is usually safe.

Many substances that produce unpleasant tastes and odors are not sufficiently volatile to be completely removed by aeration. For this reason, it is wise to consider aeration as only one of several treatments, and not to rely on it as the only treatment your water will receive.

There are many kinds of aerators on the market. One type is the spray aerator, which forces water into a high fountain. Unfortunately this kind of device isn't practical on a home scale—unless your home is Versailles or the White House. The spray requires a lot of space, and produces a spectacular show. A second drawback is that it cannot be operated during winter in cold climates.

Probably the simplest and easiest aerator is an appliance you already own—the kitchen blender. Fill it only half full with water, take the glass bubble off the top, and spin up the water.

Chemical Feeders

A wide variety of chemical solution feeders are commercially available. Whenever a chemical must be introduced into a water supply —on either a continuous or intermittent basis—it is best to supply that solution with an automatic feeder. These factors can be installed in a basement, crawl space, well pit, or even outdoors.

Feeder pumps usually are wired to operate with well pumps, so that a solution is fed at a constant ratio to the water drawn. A feeder usually introduces a solution into the water line between the well pump and the pressure tank. The pressure tank then serves as a good mixing tank and retention basin, an important function in the process of iron or hydrogen sulfide removal, or in the use of a disinfectant.

If two compatible chemicals must be added, a feed pump connected to two separate pumping heads can be used.

Chemical feeders are essential components of domestic water systems, where problems have to be treated by the homeowner. The initial cost of a feeder is high, but operation is relatively cheap. An initial trial and error period is needed to establish an exact feed rate.

Specific Treatments

Chlorine

For simple disinfection, there are several treatments available. The most common, of course, is chlorination. But water also can be disinfected with ozone, iodine, bromine, silver, and even ultraviolet light.

We have discussed the health problems associated with chlorinated water in previous chapters. Despite these problems we must recognize that chlorine is an effective killer of bacteria and viruses, and may be required for private water supplies where no previous treatment exists. Even those who use public water may have the "end of the line blues," where plenty of chlorine exists in water at the beginning of the public system, but has been used up by the end of the line.

In any case, individuals should make sure their water is free of bacteria. The danger is quite real. In fact, studies have shown the incidence of waterborne diseases is on the rise again, after many years of decline[32] (see Figure 7-9).

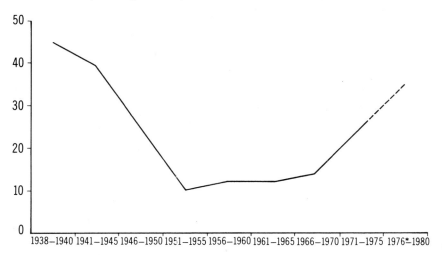

*Number Outbreaks 1976–1977

Figure 7-9: Average Annual Number of Waterborne Disease Outbreaks, 1938–1977.

(Source: Foodborne and Waterborne Disease Outbreaks Annual Summary 1977 *[Atlanta, GA: Center for Disease Control, 1979], p. 69.)*

If you decide to chlorinate your water, a system should be developed whereby chlorine will kill germs and bacteria, but then be removed from the water before you drink it. This system provides chlorine's benefits without its related problems. Not only will chlorine improve the quality of your water, it will also improve the operation of any filter you use after chlorination by reducing bacteria and algae that can grow in filters.[33]

In public water treatment plants, chlorine is measured in two ways. First, enough chlorine is added to destroy unwanted organisms. Chlorine is like money in your pocket—everytime it kills an organism a portion of it is spent. Assuming the exact quantity of chlorine is present in the water to kill all the organisms, all of it will be used up by the time it's finished doing its job.

In public water systems, enough chlorine is added to water to kill germs and viruses. But an additional amount is added: residual chlorine. It is intended to remain in the water as it passes through the mains and service lines, killing any newly introduced bacteria as it travels to your home.[34]

In your own home system, you do not need any residual chlorine. Just add enough to kill dangerous organisms, then filter out any that's left—leaving the water pure and safe.

You can buy or make a small mechanical unit called a chlorinator, to add chlorine to your water. The amount of chlorine you need depends on how bad your water is. The chlorinator should be attached to the pipe that leads to your water storage tank, because it is vital the chlorine be introduced when water is moving at a steady rate. The chlorinated water can then be stored, and as demand requires it will be withdrawn from the tank, and passed through a filter.

Here's how to test for the amount of chlorine your supply requires. You will need the following materials: a clean glass jar with a 500-milliliter capacity or more, a stock chlorine solution containing 1 percent available chlorine, dropping pipette calibrated to deliver 20 drops per milliliter, means to measure residual chlorine,[35] and a Chlorotex reagent with matching card—or residual chlorine test kit from Hack Chemical Co., Ames, Iowa.

The chlorine solution can be prepared from either dry chloride of lime (35 percent available chlorine) or high-test hypochlorite (70 percent available chlorine), dissolved in clean water. The solution should be kept in a dark bottle with a glass stopper, but it should not be kept for any length of time because it loses its effectiveness rather quickly.

To Test

Measure 500 milliliters of the sample into a clean glass jar and add 3 drops (0.15 milliliter) of stock chlorine solution. Mix and allow to stand for thirty minutes in a dark place, so the chlorine can act on the water.

Next, find out how much chlorine is left in the water after it has done its job. What's left is called residual chlorine. You can find that amount of chlorine using a test kit from the Hack Chemical Company or using the BDH Chlorotex reagent with matching card. Follow the directions provided with the kit.

Either test will show that a color is present and develop immediately when the testing agent is added to the water. A card provided with each kit will show exactly how much chlorine is left over.

Finally, subtract the amount remaining from the amount of chlorine first added. This number tells you your chlorine requirement. For example, you added 3 drops of chlorine to 500 milliliters. If 1 drop remains, your water requires 2 drops per 500 milliliters to effectively kill bacteria.

If the sample is heavily polluted, you may find no residual chlorine. In that case, begin again using a higher chlorine dose. Six drops should work well.

How It Works

You can use liquid bleach, or dry chlorine powder. Chlorinated lime, a powder, contains 35 percent available chlorine by weight. Thus 1 pound of powder to 100,000 gallons will give a dose of 0.35 milliliter/liter.

A typical chlorinator is a small mechanical unit that injects a measured amount of chlorine solution into the pipe line whenever the water pump is working.

It is possible to make your own chlorinator. You will need a large metal drum which has been painted inside with bituminous paint to protect it from corrosion. In the bottom of the drum, insert a small tube to which a clamp has been attached. By adjusting the clamp, you can control the amount of chlorine solution entering your water system.

If chloride of lime is used, you will have to stir the solution regularly. If you use household bleach, you will be spared this job.

To ensure effectiveness, chlorine must be in contact with water for thirty minutes or more.[36]

Commercial chlorinators are not overly expensive, and probably worth the money they cost to ensure you of a steady flow rate. Normally, they work when the water pump runs.

Table 7-2: Chlorination Scorecard

Chlorination vs. Bacteria	effective[1]
Chlorination vs. Viruses	effective with most viruses, but complete effectiveness not yet determined[2]
Speed of Kill	requires 30 min[3]
Effect of Minerals in Water	some chlorine is "used up" by iron and sulfur; dosage may have to be adjusted for these elements[4]
Effect of Incoming Water Temperature	as temperature goes up, the faster chlorine works[5]
Effect of Suspended Particles	some may protect viruses[6]
Advantages in Addition to Disinfection	can be used to remove sulfur, iron, and certain tastes and odors[7]; kills iron and sulfur bacteria[8]

Sources:
1. Safe Drinking Water Committee, National Research Council: *Drinking Water and Health* (Washington, DC: National Academy of Sciences, 1977), p. 119.
2. Ibid., p. 108.
3. Ibid.
4. Agricultural Research Service, Farmers' Bulletin No. 2248, *Treating Farmstead and Rural Home Water Systems* (Washington, DC: U.S. Department of Agriculture, 1977), p. 5.
5. Office of Water Programs, U.S. Environmental Protection Agency, *Manual of Individual Water Supply Systems* (Washington, DC: U.S. Environmental Protection Agency, 1974), p. 77.
6. *Drinking Water and Health*, p. 186.
7. *Treating Farmstead and Rural Home Water Systems*, p. 4.
8. *Manual of Individual Water Supply Systems*, p. 84.

Ultraviolet Radiation

It is possible to disinfect water by exposing it to a quartz–mercury vapor lamp that emits germicidal ultraviolet radiation. Some public waterworks in Germany and Great Britain use this method to treat city supplies, and it has been rumored that this method is very popular in the Soviet Union.

For home use, several manufacturers offer ultraviolet lamps that can be installed into a simple system. The procedure is very easy. All you need is a unit that gives out ultraviolet rays of the proper wavelength, and a detector that will constantly monitor the dosage.

Ultraviolet radiation is a good treatment because it does not introduce anything into the water—no chemicals, odors, fumes, or tastes. Only a short contact time is required for a UV unit to do its job. And if you should accidentally over-radiate the water, no harm is done.

But on the dark side, certain spores and viruses are fairly resistant to UV treatment, the equipment and the energy used by the unit are expensive, and frequent maintenance is required by some units.[37]

A home unit consists of one or more germicidal lamps, or ultraviolet tubes, sealed inside a narrow steel cylinder. The cylinder is connected to the supply line so that the untreated water goes into one end of the cylinder, passes over the UV lamp, and then goes out the opposite end into the water line. Radiation kills the bacteria and makes the water safe to drink. Before installation of a UV purifier, you have to know the minimum UV transmission of your water source, as well as its total bacterial count and coliform count to ensure that a specific model will provide the correct dosage. You should also know if the quality of your water fluctuates greatly. Well water generally remains consistent, but the quality of surface water often changes with the seasons.

Ultraviolet rays are not fully effective in cloudy or dirty water. Moreover, the unit can become coated and discolored, and will have to be removed and cleaned. Some experts suggest the installation of a sediment filter in the line just ahead of the UV unit to trap tiny floating particles that can influence the unit's effectiveness.[38]

UV units may not be best for water with a high iron content, because UV converts the soluble ferrous form to the less soluble ferric form, which will deposit on the tube jacket, preventing transmission of ultraviolet light.[39]

While dissolved iron and certain other dissolved organics (phenol, humic acids, lignin) have an adverse effect on the efficiency of the UV unit, the operation is not affected by pH or temperature. It does not leave any residue, and it works quickly.[40]

The units are expensive, ranging from $300 to $600, with the price varying with the treatment capacity. Most UV bulbs are easy to replace. When a light burns out, it can be replaced like any fluorescent tube.

Sources for ultraviolet units include:

> Trojan Environmental Products Inc.
> P.O. Box 2341
> London, Ontario N6A 4G3, Canada

> Aquafine Corporation
> 1869 Victory Place
> Burbank, CA 91504

> Sanitron
> Atlantic Ultraviolet
> 250 North Fehr Way
> Bay Shore, NY 11706

Ozone

Ozone has been used in Europe very effectively to disinfect water. In fact, the city of Paris, France, now has such hygienic water, thanks to ozone, there's no excuse to drink wine anymore.

The word ozone comes from the Greek word meaning smell. And smell it does! It is a pale blue, unstable gas. It is also a form of oxygen, discovered by a Dutch scientist in the late eighteenth century.[41]

Despite its popularity in Europe, especially in France, ozonation never really caught on in the United States. While it is an excellent germicide, it has certain disadvantages. It is expensive, it must be produced electrically as needed, and it cannot be stored.

Ozone is made by passing dry air between two high-potential electrodes to convert oxygen into ozone. This gas has been found to be many more times effective than chlorine in killing the poliomyelitis virus, and several other hard-to-kill organisms.[42]

The cost of using ozone, at this time, is ten to fifteen times the cost of chlorination.[43]

While ozonation is effective, it is expensive. Home technology is almost nonexistent.

Iodine

Iodine is effective in destroying bacteria, viruses, cysts, and other contaminants in water. Often it is used by the military for disinfection in the field, and by campers and travelers. Iodine should be considered a short-term means of disinfecting water. It is not recommended for use by pregnant women, and not considered safe for long-term use.[44]

Bromine

Although bromine is a good germicide, there are no reports of it being used to clean public water supplies. To date, it has been used mostly in swimming pools and by industry.

As a swimming pool disinfectant, it appears to be superior to chlorine. It is more stable during storage, produces less odor, and it does not cause "swimmer's red eye" the way chlorine does.

Chances are slim that bromine will be developed on either a public or private level, because the cost of this germicide is greater than the cost of chlorine.[45]

Corrosion Control

Corrosive water can destroy a water heater and the rest of your plumbing in quick order. Correcting the problem is not always easy because there are many causes of corrosion, and the exact cause must be found before the right treatment can be employed.

Generally, water is corrosive if it is acid. A pH measurement of 7 or lower means your water is acid to some degree. The lower the pH number, the more acid the water is.

To control corrosion, the pH of water can be raised by installing a pH adjustment unit into your home water system. This unit can add potassium carbonate or potassium hydroxide, a substance with such a high pH it not only neutralizes the acid in the water, but also kills bacteria, viruses, and other microorganisms. Moreover, since potassium is frequently in short supply in our diet, it can be beneficial to health.

Many people use soda ash or caustic soda to fight corrosion. Despite its common use, we do not recommend these treatments. Soda ash can introduce undesirable and unhealthy amounts of sodium to the water you drink. Caustic soda poses a different health problem. If the unit malfunctions, or if you misjudge the amount to add, a high caustic soda content in the water can result in severe burns.

If corrosion is caused by *high alkalinity* rather than acidity, it may be best to run the water through a water softener or anion-exchange medium. The softener will remove most of the alkalinity as it softens. The anion-exchange resin will remove substantially all the anions (carbonates, bicarbonates, and sulfates) and replace them with an equivalent amount of chlorides. The addition of these chlorides will make the water unfit for drinking. However, this treatment is a necessary preliminary to reverse osmosis treatment for water with a high pH. Reverse osmosis will then remove the chlorides, making the water acceptable for drinking as well as beneficial to household plumbing.

Iron and Manganese Removal

If water has iron in concentrations greater than 0.3 part per million, it is not considered satisfactory. While most water contains less than 5 parts per million of iron, it is not rare to find water containing 15 parts per million of iron. Extreme conditions can produce water with as much as 60 parts per million.

Manganese is often found together with iron, though in much smaller quantities. Treatment for one will remove the other. For that reason these two elements will be considered as a single problem.

The method for iron removal depends on several factors: what form the iron is in, the quantity present, and other water quality problems that might be treated at the same time.

There are three forms of iron commonly found in water:

Dissolved iron is present when the water is clear and colorless at first, but forms rust particles as it is exposed to air. Manganese acts in a similar way when exposed to air, but forms a brownish black precipitate.

Oxidized iron is commonly called rust. Rust in water means that the iron content has been exposed to air or another oxidizing agent.

Bacterial iron is a third form of iron, created by iron bacteria. Water with iron bacteria will form a black slime on the surface of any place it's stored . . . like the walls of a toilet tank, the inside of a hot water heater, or the top of a water softener. Iron bacteria really foul up a water system with ugly slime.[46]

There are several ways to control these different forms of iron and manganese. In theory, the techniques are simple. But in fact, they're pretty tricky because you have to know just what type of iron is in your water before you can treat it properly.

Ask yourself the following questions to help you determine the source of the iron and manganese in your water.

• Is your water corrosive? Does it have a low pH? It is possible the iron in your water is simply leaching from the household plumbing and does not exist in the raw water at all. If this is the case, your water should be treated for its corrosiveness.

• Do you have that telltale ring around the toilet bowl? Do your clean clothes have rusty blotches on them? Chances are you are afflicted with dissolved iron.

• Is there a black slime inside your toilet tank? Do brown or black flecks appear in your water? It looks like iron bacteria have taken up residence in your home.

To be absolutely sure what kind of iron problem you have, have a test made before you spend any money on treatment. The test will save you money in the long run.

Ion Exchange

Most regular water softeners can remove iron and manganese just as they remove calcium and magnesium. However, only iron in a dissolved form can be removed. The chemical process is the same as that used for water softening—iron ions are replaced by sodium ions.

The amount of iron that can be removed by a water softener depends on the exchange material used in the unit, the design of the softener, and other variables—like whether the water is hard.

Some water conditioning experts say the capacity for iron removal is equal to the capacity for removal of calcium and magnesium, if these hardness minerals are not present. Others feel the limits for iron are much lower. Some manufacturers recommend using a softener for iron treatment only when the iron level is below 2 parts per million. Still others claim their units can remove iron from water containing up to 10 parts per million of dissolved iron.[47]

Before you rely on your water softener to do the job, be sure the iron in your water is the dissolved kind. Then ask your dealer about the specific capacity and efficiency of your home unit.

Some disadvantages of using a softener are that iron particles can become lodged in the exchange bed. If the softener uses zeolite, backwashing the softener can be difficult because the particles are heavier than zeolite. If the softener uses resin beads, they can become coated by a heavy, gelatinous substance called ferric hydroxide. Should this situation develop, contact the manufacturer for cleaning directions.

Stabilization with Polyphosphates

This procedure removes only dissolved iron. Polyphosphates do not actually remove iron from water. Instead, they stabilize and disperse the iron so that it cannot be oxidized when exposed to air. Chemists speak of the iron as being sequestered. The polyphosphates tie up the dissolved iron, inhibiting its normal oxidation. The water will stay clear even when exposed to air, and it will not stain fixtures or clothes.[48]

Polyphosphates are often used in treatment plants to soften water chemically by sequestering hardness minerals, and to control corrosion by depositing a thick film on the interior of water pipes, heaters, and pressure tanks.

For this home treatment, you will need food-grade phosphate sold under the names Zeotone, Micromet, or Nalco M-1. The chemical is added to the water with a chemical feeder before it reaches the pressure tank; the air in the tank will cause rust to form before the polyphosphates can work.

This treatment will inactivate between 1 and 2 parts per million of dissolved iron.

The treatment is not without drawbacks. For example, heat can reduce the effect of the polyphosphates.[49] Cooking, brewing coffee or tea, or heating water for any length of time can trigger a chemical change so that the polyphosphate no longer controls iron.

This treatment is often recommended for installation with a water softener because it protects the softener from becoming clogged with iron.

Maintenance of the chemical feeder required for this treatment involves keeping the feeder supplied with the polyphosphate material.

Most feeders are big enough to hold about a thirty-day supply.

Polyphosphate treatment will be effective in a pH range of 5.0 to 8.0. All other iron removal techniques are designed to treat iron waters with essentially a neutral pH.

If the pH of the water is lower than 5.0, an alkali such as potassium carbonate or potassium hydroxide may be used to raise the pH, followed by chlorine or another disinfectant for iron precipitation. A simple filter could then remove the precipitated iron.

Precipitation and Filtration to Remove Iron

Most techniques to remove iron and manganese from water depend on some type of precipitation and filtration. Because iron generally exists in both soluble and insoluble forms, this is probably the most thorough treatment of all since it removes both types of iron.

Precipitation and filtration oxidize iron and manganese and they become insoluble. The resultant rust then can be settled out or filtered out of the water.[50]

Chlorine can be used to oxidize iron, especially iron that is bound to organic compounds. Chlorine disinfects at the same time and kills any iron bacteria along with any other microorganisms.[51]

But chlorine isn't the only oxidizing agent. Potassium permanganate and chlorine dioxide also induce precipitation. These strong oxidants are employed chiefly where water contains manganese.[52] Lime is used as an acid in water where pH is too low to allow rapid oxidation.

Once iron is oxidized, it is in a form that will settle out of the water. The next step is to store the water in retention basins for twenty or thirty minutes while the rust particles drop to the bottom. Depending on the volume of water to be treated, these retention basins might have to be quite large. In fact, this system may be practical only for farms.

Once the rust has settled to the bottom of the holding basins, the water is drawn off the top and filtered to remove all insoluble iron and manganese. A fine material like crushed anthracite coal, diatomaceous earth, or activated carbon can be used. An activated carbon filter is best because it gives the added advantage of removing tastes, odors, colors, excess chlorine, and other chlorinated compounds. It will also remove any unreacted potassium permanganate, if that chemical was used in the oxidation procedure.

Oxidizing Filters

These filters convert dissolved iron into insoluble iron by passing the iron-heavy water through a filter made of material treated with manganese. As the iron forms rust particles, it is filtered out by a granular material in the filter. Any iron that oxidized before it got to the filter is also strained out.

While these filters offer a convenient, one-step way to get rid of iron, they do not work on iron bacteria. In fact, bacterial slime clogs these filters and mineral beds, making filtration impossible.

Another drawback is that oxidizing filters must be backwashed and rinsed every week to remove precipitated iron from the filter bed. This regular maintenance is essential to keep accumulated rust particles from entering the finished water supply as "slugs" of iron during periods of heavy usage.

When iron accumulation becomes excessive, regeneration will be necessary as often as every second or third time the filter is backwashed. Frequency of regeneration is determined by the amount of iron and manganese in the water, the amount of hydrogen sulfide in the water, the size of the filter, the specific mineral used, and the amount of water used.

Regeneration consists of passing a strong oxidating agent (usually potassium permanganate) through the filter after it is backwashed. It will reoxidize and restore the manganese dioxide coating on the filter.

Control of Hydrogen Sulfide

The treatment methods used for iron removal are suitable for hydrogen sulfide removal. When the gas is oxidized, elemental sulfur and water are produced. The sulfur forms an insoluble yellow powder, which then can be removed by filtration. The same methods used to oxidize dissolved iron will oxidize hydrogen sulfide.

Aeration or an oxidizing filter using manganese dioxide will be effective for removing low to moderate amounts of hydrogen sulfide. But when levels exceed 10 parts per million, the oxidation capacity of these filters is surpassed. Then, a stronger oxidation agent like chlorine or potassium permanganate is needed. Normally, a chemical feeder is required for this treatment. The chlorine will also provide a residual to retard the growth of sulfate-reducing bacteria in the pipes, pressure tank, and water heater. As in the case of iron, the solution should enter the water before the pressure tank in order to provide adequate contact time. This tank serves as both a contact and retention tank, and it should be large enough to retain the water long enough. For example, if there are 20 parts per million hydrogen sulfide in the water and only enough contact time provided to remove 15 parts per million, the tank is large enough.

Use of iron removal techniques to treat for hydrogen sulfide has several advantages. The two often are found together in well water and so common treatment methods are convenient. Because the same chemicals, same methods, and same filters can be used, it saves money to double up. However, treatment for iron and hydrogen sulfide together will compromise the capacity of the filter to remove iron alone if hydrogen sulfide is present since some of the filter's potential to oxidize the iron will be spent on the hydrogen sulfide. Generally, hydrogen sulfide has twice the oxidation demand of iron.

How to Determine Filter Regeneration Needs

Suppose the filter you are using has a rated capacity of 5,000 ppm-gallons for hydrogen sulfide removal if no iron is present, and 10,000 ppm-gallons for iron removal if no hydrogen sulfide is present. If both iron and hydrogen sulfide are present, the capacity is compromised. Your family of four uses well water containing 3 ppm hydrogen sulfide and 1.5 ppm iron. There are no iron or sulfur bacteria present. To determine the total oxidation demand in terms of iron, multiply the 3 ppm hydrogen sulfide by 2 to get 6 ppm iron. Add this to the iron in the water to get 7.5 ppm iron. This means that your water has an oxidation demand equivalent to 7.5 ppm of iron. Your family uses 50 gallons of water per person per day, or 200 gallons per day (gpd). Multiplying this by your oxidation demand, or 7.5 ppm times 200 gpd, gives 1,500 ppm-gpd. This tells you that every day you will use up 1,500 ppm-gallons of your filter's capacity to remove iron. With your filter rated at 10,000 ppm-gallons, the filter will need to be regenerated every 6.6 days, or to be safe, every 6 days.

Some dealers provide a service whereby they come and replace a spent filter with a new one that has been regenerated at the factory.

Oxidizing filters generally have strict requirements for pH levels, and maximum amounts of hydrogen sulfide and iron. Some also have a minimum requirement for silica in the water.

After water has been treated to remove hydrogen sulfide, be sure it is treated again—perhaps with a GAC filter or RO unit beneath your kitchen sink—to remove any remaining chlorine.

NOTES

1. *Drinking Water and Health* (Washington, DC: National Academy of Sciences, 1977), p. 1.

2. Donald E. Carr, *Death of the Sweet Waters* (New York: W. W. Norton and Company, 1971), p. 28.

3. *Drinking Water and Health*, p. 11.

4. William D. G. Murray, *Important Characteristics to Look for in a Water Purifier* (Malvern, PA: General Ecology, Inc., 1977).

5. American Water Works Association, Inc., *Water Quality and Treatment* (New York: McGraw-Hill Book Co., 1971), p. 243.

6. Ibid., p. 244.

7. Office of Water Programs, U.S. Environmental Protection Agency, *Manual of Individual Water Supply Systems* (Washington, DC: U.S. Environmental Protection Agency, 1974), p. 75.

8. Agricultural Research Service, Farmers' Bulletin No. 2248, *Treating Farmstead and Rural Home Water Systems* (Washington, DC: U.S. Department of Agriculture, 1977), p. 15.

9. August B. Russo, "Install a Filter and Clean Up Your Home Water Supply: Here's the Quick Way to Improve the Look, Smell and Taste of Your Drinking Water," *Popular Mechanics,* May, 1978, pp. 136–137.

10. Lucia Mouat, "Faucet Filters: Mostly a Waste of Money: And a Neglected One Can Be a Polluter," *Christian Science Monitor,* March 28, 1979, p. 14.

11. Herbert Shuldiner, "The Truth About Clean-Water Devices," *Popular Science,* May, 1975, p. 48.

12. Ibid., p. 50.

13. "Those Gadgets that Make Water 'Pure,'" *Changing Times, The Kiplinger Magazine,* February, 1978, p. 40.

14. Shuldiner, "The Truth About Clean-Water Devices," p. 46.

15. *Development of Basic Data and Knowledge Regarding Organic Removal Capabilities of Commercially Available Home Water Treatment Units Utilizing Activated Carbon: Preliminary Report—Phase 1* (New Orleans: Gulf South Research Institute, 1979), p. 17.

16. Ibid., p. iii.

17. Ibid., p. 4.

18. Ibid., p. 7.

19. Ibid., p. 36.

20. Ibid., p. 5.

21. Ibid., p. 35.

22. Thomas J. Basalik, *Reagent Grade Water for Laboratories* (Boston: Barnstead Co., 1978).

23. *Water Quality and Treatment,* p. 342.

24. Dean Spatz, "Reverse Osmosis Reclamation Systems for the Plater," *Finishers' Management,* July, 1971.

25. George H. Klumb, *Reverse Osmosis—A New Principle in Water Purification* (Northbrook, IL: Culligan International Company).

26. George H. Klumb, "Reverse Osmosis—An Important Process in the Purification of Water for Parenteral Administration," *Bulletin of the Parenteral Drug Association,* October, 1975, pp. 261–268.

27. Ibid.

28. Ibid.

29. Spatz, "Reverse Osmosis Reclamation Systems."

30. Klumb, "Reverse Osmosis," p. 267.

31. *Manual of Individual Water Supply Systems,* pp. 90–91.

32. Gordon L. Culp and Russell L. Culp, *New Concepts in Water Purification* (New York: Van Nostrand Reinhold Co., 1974), p. 179.

33. *Water Quality and Treatment,* p. 203.

34. Ibid., p. 207.

35. These items can be purchased from a laboratory supply house.

36. *Water Quality and Treatment,* p. 215.

37. Ibid., p. 164.

38. N. Henry Wooding, Jr., *Make Your Water Supply Safe* (Harrisburg, PA: Department of Environmental Resources), p. 6.

39. R. W. Yip and D. E. Konasewich, "Ultraviolet Sterilization of Water—Its Potential and Limitations," *Water and Pollution Control,* June, 1972, p. 17.

40. Ibid.

41. *Water Quality and Treatment,* p. 173.

42. Culp, *New Concepts in Water Purification,* p. 212.

43. *Water Quality and Treatment,* p. 175.

44. Wooding, Jr., *Make Your Water Supply Safe,* p. 6.

45. Culp, *New Concepts in Water Purification,* p. 212.

46. *Manual of Individual Water Supply Systems,* pp. 83–84.

47. *Treating Farmstead and Rural Home Water Systems,* p. 13.

48. Ibid.

49. Ibid.

50. Ibid.

51. Ibid., p. 14.

52. *Water Quality and Treatment,* p. 387.

8 Finding a New Source

For any number of reasons, some individuals may choose to develop a new water resource rather than tackle the job of cleaning water from their old system. For those who are currently drinking water from a surface supply, this switch may be the *best* way to get good water. Surface water is notoriously polluted, and the simple change to well water—or any underground water—often can provide a dramatic improvement in water quality. Others who have badly located wells—too close to septic tanks, overflowing rivers, feed lots, and so on—may also benefit greatly by finding a new source in a cleaner spot.

Some lucky individuals may find a spring on their property. Others, especially those with a poor or unsteady water supply, may want to install a cistern to catch rainwater.

In any case, this chapter will briefly describe how to plan, develop, and maintain wells, cisterns, and springs.

Wells

If you decide to dig a well, you will not be alone. More than 700,000 new water wells are drilled in the United States each year. In fact, about 96 percent of all rural domestic supplies are derived from wells.[1]

Constructing a well is a lot more complicated than just digging a hole in the ground. And while the old oaken bucket has a certain charm, unless you have the shoulders of an Arnold Schwarzenegger, you probably will want an electric pump, selected from a confusing assortment available for water withdrawal.

The first step is finding the proper site for a well. Two things must be considered: finding a location free of contamination where water safety will be ensured, and the practicality of drilling in that spot.

A well must be located as far as possible from known or suspected sources of pollution. These sources include cesspools and septic tanks, livestock and poultry yards, milkhouse drain outlets, silo pits, landfill operations, and even floodplains of contaminated streams.

Unfortunately, nobody can give exact measurements for a safe distance between a well and a source of contamination. Topography is one variable to consider. For example, a well that is uphill from a septic tank may require a 100-foot safety zone. But the same well dug in ground that is level with a septic tank requires a greater distance.

A second variable is the type of soil in any given location.

Researchers for the water laboratory at South Carolina State College sampled water randomly drawn from three locations in that state.[2]

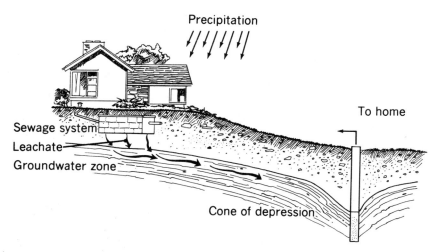

Figure 8-1: Leachate from a Septic Tank Contaminates a Nearby Well.
(Source: Is Your Drinking Water Safe? [Harrisburg, PA: Department of Environmental Resources, 1976], inside front cover.)

The samples were tested for total coliforms, *Escherichia coli*, and fecal streptococci. Their very frightening discovery was that only 7.5 percent of the water supplies tested were *not* contaminated with *E. coli*. Generally, the people using this water drank it as it came from the well, without treatment.

What researchers further found was that the number of bacteria decreased as the depth of the well increased, or as the distance between the well and septic tank increased.

Results also varied with the type of soil in each location. One area where samples were drawn is Hampton County, which is alongside the Savannah River. There the soil is generally waterlogged because of frequent river flooding. Those wet conditions seem to inhibit soil absorption, making a septic tank disposal system—where water percolates through the ground as through a filter—almost completely impractical.

A second survey site, Georgetown County, is located along the Atlantic Ocean. Soil there is sandy, allowing pollutants to rush through it, traveling great distances. Filtration improves in time, however, because a mat builds within the soil, trapping contaminants.

The study showed that pollutants can travel distances that vary greatly, depending on local soil conditions. Most experts recommend that, if soil percolation conditions are good, a minimum of 100 feet should be maintained between a water well and a waste disposal system. If the filtering properties of your soil are unknown, you can get assistance from your local health department or your state Department of Environmental Resources.

The United States Department of Agriculture offers the following suggested minimum distances (see Table 8-1).

Once you have ruled out all the unsafe well locations, look carefully at the land remaining. Before you spend a lot of money drilling holes that might come in dry, evaluate this land's water potential. Sometimes plants can point the way to the perfect well site. Look for rushes, cottontails, elderberries, reeds, cattails,[3] salt grass, pickle wood, arrow weed, palm trees, willow trees, cottonwood, mesquite, grease wood, and rabbit brush. Because these are all water-loving plants, their growth in a certain spot usually indicates the presence of water.[4]

Another way to establish a likely spot for a well is to get in touch with your friendly neighborhood dowser. Yes, that is the water witch who finds water with a forked willow branch. Today he probably uses a fiberglass rod or a slim piece of copper tubing, because the bark of

Table 8-1: Suggested Minimum Distances between Source of Contamination and Well Site

Source of Contamination	Minimum Distance (ft)
Waste disposal lagoons	300
Cesspools	150
Livestock and poultry yards	100
Privies, manure piles	100
Silo pits, seepage pits	150
Milkhouse drain outlets	100
Septic tanks and disposal fields	100
Gravity sewer or drain not pressure tight	50
Pressure-tight gravity sewer or drain	25

Source: Agricultural Research Service, Farmers' Bulletin No. 2237, *Water Supply Sources for the Farmstead and Rural Home* (Washington, DC: U.S. Department of Agriculture, 1971), p. 5.

the willow can make the dowser's palms very sore when it twists and turns.

Dowsing has been popular since at least the sixteenth century, and is still regularly practiced in many parts of this country. Moreover, it recently received a bit of scientific credence, however grudgingly. Don't feel foolish about dealing with a dowser. Various government agencies hire them to find lost water mains and new public water supplies.

Thomas Edison believed the dowsing phenomenon was related to electricity, while Einstein thought it had something to do with electromagnetism. In one study to determine the reason why dowsing works, a group of 150 volunteers at Utah State University tried their hand at the ancient art.[5] All were novices. Each was given thirty small wooden blocks and an L-shaped rod. As they walked along test paths (chosen for lack of any physical signals of water), they were asked to drop a wooden block wherever they felt a dowsing reaction. The location of the blocks was marked on a chart, and the blocks were removed before the next volunteer tried the experiment.

When all the blocks had been dropped and charted, a pattern emerged. Three out of four blocks were dropped in similar locations!

Researchers suggested that dowsing reactions may result from small magnetic field variations. When dowsing paths were measured with a device that tests for deviations in magnetic fields, some correlation with dowsing reactions was made.

Not surprisingly, a new device for finding water is being marketed by a company called Accurate Water Location, 315 New Hackensack Road, Poughkeepsie, NY 12603. The instrument is called an Aquatometer, and is a version of a nonelectronic magnetometer. It senses the presence of abnormalities in rock, and the resulting variation in the earth's magnetic field. It is a kind of robot dowser!

Water diviners claim they can locate underground rivers, lakes, or streams that will provide a plentiful supply of water for a well. But, as you already know, water does not exist beneath the earth in great rushing Hudsons and Mississippis. Rather, it is held within the spaces between gravel and rock. To get a good idea of how an aquifer looks, picture a shallow bowl filled with pebbles and sand into which water has been poured. Imagine, then, a straw inserted into this bowl through which water can be withdrawn. The bowl is the aquifer, the straw is your well.

Yet another way to establish the proper location of your well is to consult a well drilling company. Many keep careful logs of where water has been found in the past, and use these to make highly educated and usually accurate guesses about where water can be found on your property. The United States Geological Survey (Water Resources Division, National Center, Mail Stop 420, Reston, VA 22092) may also have hydrogeological maps of your area showing the known aquifers and their yields to help your guesswork along. Also state and local government agencies frequently keep files of well logs submitted by drillers under their jurisdiction; and these agencies are often helpful in evaluating the water supply potential of your area.

Once you've found the site, you have a choice of different kinds of wells. Here's a rundown of what can be constructed, and how successful each type of well is.

Dug Wells

Dug wells can be built by hand or with power tools. They are usually at least 3 feet in diameter, and most go deep enough to penetrate the water table—usually less than 50 feet.

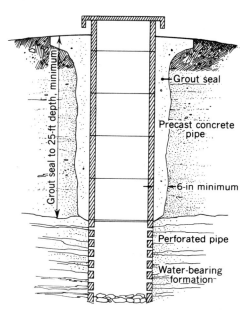

Figure 8-2: Dug Well.
(Source: Private Water Systems, *MWPS-14, Fourth Edition* [*Midwest Plan Service, Ames, IA 50011, 1979*], *p. 6, with permission of the publisher.)*

With careful construction, a dug well can be cased as tightly as any other. As the well is dug, concrete well rings can be inserted, which both seal and support the well sides. These rings are 3 feet in diameter and 2 feet in length. The well must be dug with a 4-foot diameter. After the 3-foot rings are in place, the space between the soil and the ring is filled with concrete. Below pumping level, the rings are surrounded with gravel.

The major drawback of dug wells is that they frequently bottom out during dry periods.

Bored Wells

Bored wells are similar to dug wells, but deeper and smaller in diameter. They can be constructed by hand with either manually driven or powered augers. Lengths of circular tile are inserted into the bore hole as the well is dug.

Bored wells are usually cased for their entire length. While the construction of bored wells may seem simple, one major drawback is the frequent collapse of the hole—with resulting loss of equipment.

Driven Wells

Driven wells may be the cheapest and quickest way to get water out of sandy ground. To construct a driven well, a point is forced into the ground. The point brings with it a screen. Once the screen is below the water table, coupled pipe sections can be installed as well casings.

The drivepoint, which looks a great deal like a giant masonry nail, can be pounded into the ground with a sledge hammer. Coupled to a drive pipe, with additional lengths of pipe added as needed, the point should strike water at or above 50 feet.

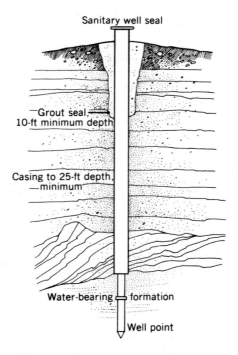

Figure 8-3: Driven Well.
(Source: Private Water Systems, *MWPS-14, Fourth Edition [Midwest Plan Service, Ames, IA 50011, 1979], p. 6, with permission of the publisher.)*

Drilled Wells

Drilled wells can go to great depths, even through tough rock formations. Because a drilled well requires sophisticated equipment and considerable training to construct, the job must be done by a professional drilling contractor. Two types of drilling methods are generally practiced today—the percussion, or cable tool method, and the rotary method (which has several variants).

A drilled well consists of a hole drilled through soil and loose rock overburden to a water-bearing formation. The upper portion is lined with well casing; the space between the outside of the casing and the overburden is filled with cement to seal off any potential avenues of contamination from the surface. If the well is completed in a sand or gravel aquifer, some sort of screen or well filter is installed below the casing. If the well is completed in bedrock, a screen is rarely needed.

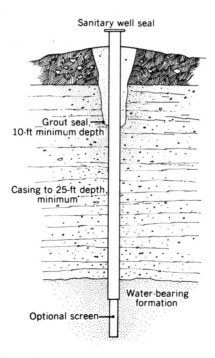

Figure 8-4: Drilled Well.
(Source: Private Water Systems, *MWPS-14, Fourth Edition [Midwest Plan Service, Ames, IA 50011, 1979], p. 6, with permission of the publisher.)*

According to statistics compiled by the National Water Well Association, the "average" domestic water well in the United States is 6 inches in diameter and around 175 feet deep. When equipped with a conventional pumping system, the total price for the system is approximately $3,000. These averages, however, should be considered in light of the wide variability in well characteristics and prices that result from the extremely diverse subsurface geology of the nation.

A call to a local drilling contractor is probably the best way to find out how deep your well will be and what it will cost.

Jetted Wells

Jetting, or hydraulic well drilling, is most successful in sandy soils. This system can be used in clay soil, but won't work in rock formations.

In jetting, the well hole is made by the force of a high velocity stream of water. The water loosens the soil and washes it upward, out of the hole. Then a bit is pushed down into the loosened soil.

Factors that determine the well's yield include the type of well, the type of aquifer, and the depth of the well's penetration into that aquifer.

Often people believe that by increasing the diameter of their well they can substantially increase its yield. However, this is not the case. Doubling the diameter will increase the yield by only about 10 percent.

Because bored and dug wells can be sunk to only a very limited depth, their yield is usually low. Driven wells can be sunk as much as 50 feet, to gain access to a good supply of water within the aquifer. However, because the diameter of a driven well is so narrow—between 1 and 2 inches—only certain types of small pumps can be used.

Drilled and jetted wells, because they are deep and can handle large pumping equipment, are the most reliable sources of a plentiful supply of water.

The yield of the well must be established before a pump can be chosen. The pump capacity cannot exceed the yield of the well or permanent damage to the pump will result.

In addition to yield, you should also consider your family's daily water needs, the size of the pressure or storage tank, the size of the well casing, the availability of power, the cost and economy of operation, the reliability of the pumping equipment, and the difference in elevation between the ground level and the water level inside the well during pumping.

For complete accuracy, you may have to depend on a professional well driller to give you this necessary information.

Once you know all the facts about your well, you can choose the right pump. There are many types available, ranging from the old hand pump that sits above the kitchen sink to a deep submersible pump with a pitless adapter for easy maintenance.

Selections include: reciprocating pumps, centrifugal pumps, submersible turbines, jet pumps, and rotary pumps. Each is adaptable

Table 8-2: Wells Suitable for Differing Geologic Formations

Type of Well	Depth	Diameter	Geologic Formation
Dug	0–50 ft	3–20 ft	suitable: clay, silt, sand, gravel, cemented gravel, boulders, soft sandstone, and soft, fractured limestone unsuitable: dense igneous rock
Bored	0–100 ft	2–30 in	suitable: clay, silt, sand, gravel, boulders less than well diameter, soft sandstone, soft, fractured limestone unsuitable: dense igneous rock
Driven	0–50 ft	1¼–2 in	suitable: clay, silt, sand, fine gravel, and sandstone in thin layers unsuitable: cemented gravel, boulders, limestone, dense igneous rock
Drilled (cable tool)	0–1,000 ft	4–18 in	suitable: clay, silt, sand, gravel, cemented gravel, boulders (in firm bedding), sandstone, limestone, and dense igneous rock
Rotary	0–1,000 ft	4–24 in	suitable: clay, silt, sand, gravel, cemented gravel, boulders (difficult), sandstone, limestone, and dense igneous rock
Jetted	0–1,000 ft	4–12 in	suitable: clay, silt, sand, ¼-in pea gravel unsuitable: cemented gravel, boulders, sandstone, limestone, and dense igneous rock

Source: Agricultural Research Service, Farmers' Bulletin No. 2237, *Water Supply Sources for the Farmstead and Rural Home* (Washington, DC: U.S. Department of Agriculture, 1971), p. 7.

to both shallow and deep wells. You will have to discuss the specifics of your well with a well driller to be sure you purchase the correct pump.

Well Water Protection

Once the well is constructed, it must be protected from contamination during use. First and foremost, the well should be sealed at the top to keep out surface water, runoff, debris, and even small animals. Should contaminants get into the well and mix with your drinking water, the well will have to be cleaned and disinfected.

The seal should be watertight, and raised at least 2 feet above the highest known flood level in your area. A cement slab is not sufficient protection because water and burrowing animals can easily get under it and into the well. Some seal designs are available that employ a gasket held between two steel plates.

Any seal should be accessible and easily removed so the well can be inspected, serviced, and maintained in good condition.

Once the well has been constructed and the pump installed, the well has to be flushed to remove rock dust and other construction debris. Next, it is vital to disinfect the entire water supply system. Bacteria introduced into the system during construction—or after repair of an existing well—can be harmful. Generally, disinfection is done by the well drilling company and often is routinely included in the contract. If you must disinfect on your own, here's the procedure according to the United States Department of Agriculture.[6]

Wait until evening, because water cannot be used during the eight- to twelve-hour disinfection period.

Open the well by removing the cap or sanitary seal and pour in 1 gallon of chlorine bleach.

Connect a garden hose to the nearest hose bib, and draw water through it until there is a strong odor of chlorine. Wash the inside of the well casing or walls with the chlorine solution. Then wash the well cap or well seal.

Inside the house, allow water to flow from each faucet—one at a time—until you smell chlorine. Once the odor is detected, turn off the faucet and move on to the next one. If you cannot smell chlorine, begin again by adding more chlorine bleach and washing the well's innards.

Let the water stand in the pipes at least eight hours, and up to twelve hours. Do not use this chlorine water for any reason. In the morning, let the water run from all faucets until you no longer can smell chlorine. However, be cautious about disposal, as heavily chlorinated water may kill grass or shrubbery.

This disinfection procedure also can be used for old wells that have become contaminated. It will effectively clean the well of bacteria provided the source of contamination has been eliminated.

If you are abandoning your old well in favor of a new one, *never* use it for waste disposal. It will contaminate your aquifer. Regulations may exist in your state requiring you to fill an abandoned well. Often concrete is used to ensure permanent sealing.

Cisterns

A cistern is nothing more than a big tank—usually buried underground—that stores rainwater.

Some people swear that cisterns provide perfectly good drinking water, while others vigorously argue that such water is fit only for washing cars or sprinkling lawns. Evidently, the proponents have

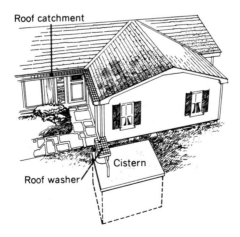

Figure 8-5: Roof Catchment and Cistern.
(Source: Private Water Systems, *MWPS-14, Fourth Edition [Midwest Plan Service, Ames, IA 50011, 1979], p. 12, with permission of the publisher.)*

sipped the water from a clean, sealed, and properly maintained cistern. Most likely, the opponents have tasted the slimy concoction that stagnates in old, unclean tanks.

The old practice of collecting rainwater from the roof and storing it in an underground cistern is regaining popularity, and is often considered a practical source of additional water to supplement a meager home supply. The deteriorating quality of tap water and severe drought have created new interest in cisterns even among urban dwellers.

Most people do not use this water for drinking. Because it is soft, they prefer to save it for washing and laundering, and even for hot water heating. While it is true that rainwater is acid and can corrode plumbing, the acidity can be corrected inside the cistern, before it enters the household system.

Other people do use cistern water for drinking. In fact, in some rural areas, cisterns provide the sole source of water. In the Shenandoah Valley, for example, about half the homes built before 1960 have cisterns. Lacking other water sources, runoff from the roofs of farm buildings may be able to provide you with an inexpensive supply of acceptable water.

Acid Rainfall and Pollution from Roof Surfaces

Rainwater itself is not pure water, contrary to what most people think. As it falls, it scrubs the air, absorbing various air pollutants. (See Chapter 3.)

The most important characteristic of rainwater in a cistern is its acidity. The more acid the water is, the faster it will leach heavy metals and organic compounds from the surfaces it touches. The type of roof this water falls on is as important as the stone or cement in the cistern that holds it.

Slate shingles and terra-cotta tile, used as roofing on many older houses, are very good roofing materials for catching rainwater. But some other roofing materials contaminate the rainwater that runs off them.

Galvanized tin roofs are quite common on farm buildings. Unfortunately, this roofing material is very susceptible to leaching of heavy metals by rainwater. The more acidic the rainfall, the more these roofs will rust—and zinc and cadmium will be among the impurities in the runoff.

This leaching can be stopped by painting the roof. Be sure to use only lead-free paint, because lead can be leached just as easily as the zinc and cadmium.

If water is collected from a roof covered with asbestos shingles, it may contain asbestos particles that pose a serious health threat, making the water unfit for drinking until it has been tested and properly treated. A reverse osmosis unit can efficiently remove asbestos particles from water.

The first rainwater that runs off the roof washes away the soot, dust particles, and bird droppings that have accumulated since the last rainfall. A roof washer or cutoff valve can prevent most of these contaminants from getting into your water supply.

Before you drink *any* cistern water, be sure to have it tested and treated as needed for bacteria, as well as the heavy metals and asbestos particles that may be washed off from the roof by the rain. If you find the water is unsatisfactory, treat it as you would well water or any other supply. It can be chlorinated and filtered through granular activated carbon, or treated with any of the methods described in Chapter 7. If you have no other option than to use a cistern as your primary source of drinking water, you should consider having it tested on a regular basis —at least once a year—as an important part of proper maintenance.

Careful Construction for High-Quality Water

There are several features characterizing a clean, good cistern. Smooth, watertight walls prevent groundwater from getting in and cistern water from leaking out. The inlet and overflow spouts should be covered with fine screening. A home with a cistern should have a roof washer or cutoff valve, both devices that keep the first rainwater out of the cistern. In addition, a good cistern has an entrance—usually a manhole—to allow someone inside to clean it, and a tight-fitting manhole cover. The ground around a cistern usually is sloped so that surface water cannot enter. Finally, a cistern should be built to hold enough water to meet a family's needs during a drought.

To keep the first part of the rainfall from entering the cistern and polluting the cleaner water with dirt from the roof, you need a way to divert the first several gallons.

Some people install cutoff valves in their downspouting for this purpose. The valve has a lever that can be set to allow water to enter the cistern, or switched to divert it elsewhere. It is a simple valve, easy to install and to work. But it has one major drawback. Someone must be home when it rains, so the lever can be adjusted to let water into the cistern once the roof is clean. And someone must be home again when the rain stops, to flip the lever back to its cutoff position.

However, you can install a filter to provide some protection for those times when the lever has not been turned to its cutoff position and the rains have come. These filters will keep the worst dirt from your roof out of your cistern.

A downspout filter is made from a container holding a base layer of 6 to 8 inches of gravel, topped by several inches of sand. Some people improve on these filters by adding activated carbon; others mistakenly use ordinary barbecue charcoal, which can do more harm than good. The sand in these filters acts as a fine screen as well as a bacterial filter. But these filters must be backwashed and cleaned frequently. The top layer of sand has to be replaced often or bacteria will infest it. Without scrupulous and frequent maintenance, these filters can create more pollution than they remove.

A better choice for most people, then, is to employ a roof washer. You can buy one or you can make your own.

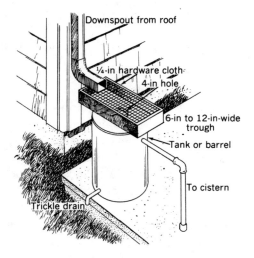

Figure 8-6: Homemade Roof Washer.
(Adapted from Private Water Systems, *MWPS-14, Fourth Edition [Midwest Plan Service, Ames, IA 50011, 1979], p. 14, with permission of the publisher.)*

To make one, you'll need a simple tank, barrel, or garbage can. Connect it to the downspouting where it will catch the dirty water from the first part of the rain. After the container has filled, water will overflow into a pipe that is connected to the cistern. Many people install a drain faucet at the bottom of this tank. Left slightly open at all times, it works as an automatic drain and also prevents the water from freezing when the temperature is low. An added benefit is that, with the faucet wide open, manual cleaning of the barrel is made easy.

Another protective device is common screening. Gutters should be screened to keep out leaves and twigs, but installed so they can be removed easily for cleaning. Also put screens at the top of downspouts so leaves can't clog them.

Cleaning a Cistern

Rarely does a health department or extension service provide instructions for cistern maintenance. Instead, people are left to their own devices, and generally rely on advice from neighbors or friends. The advice can vary greatly. Some people clean their cisterns annually, while others wait for five years. Cleaning agents, too, vary from person to person.

Here's how to do the job right. First, pump the cistern nearly dry. Wearing boots and carrying a flashlight, climb inside. Inspect the walls for cracks or leaks. These must be patched with cement before you can go further.

Next, begin to scrub the inside walls. You can make a cleaning solution from any of the following mixes:

- 3 parts vinegar to 1 part water
- 2 pounds of baking soda dissolved in 2 gallons of water
- 4 pounds of zinc sulfate mixed with 1 gallon of water

The zinc sulfate solution requires three to four hours to work before the cistern can be flushed out. The baking soda and vinegar solutions need thirty-six hours to do their job. The cleaning solution must be removed by washing the cistern walls with a hose, and then pumping the drain water out of the cistern.

The job sounds easier than it is. The work is hard labor, intensified by the lack of fresh air. One small help is to direct a fan into the cistern while you are working.

After the cistern has been cleaned, many people whitewash the internal walls. We don't advise this procedure because prolonged storage will dissolve some of the alkaline lime into the water, changing it from soft to slightly hard. Instead, we recommend disinfecting the tank by spraying a chlorine solution on the internal walls. Do not enter the cistern after this disinfection. Do not allow water to reenter the cistern until the chlorine has dissipated.

Determining the Cistern's Size

Several factors have to be considered in deciding the size of the cistern to be built. Determine how much water is used each day, the amount of water storage needed and how long that will last, the size of the building(s) guttered into the cistern, and the amount of rainfall in your area.

For example, the amount of water needed by a family can be calculated from the amount used by each person per day—usually between 50 and 80 gallons—and multiplied by the number of people in the family. Thus a family of four will need a minimum of 200 gallons each day, or a maximum of 320.

To provide sufficient water for a dry period, you must multiply the family's daily need by the number of days in an average dry spell. In the East, parts of the Midwest and South, that average is ninety days. Therefore, a family of four requiring 200 gallons of water a day for a ninety-day period must have a cistern capable of holding 18,000 gallons as a minimum.

To determine how much water a cistern will supply, you need additional critical information. What is the minimum yearly rainfall recorded in your area? The longest period of drought recorded in your area? What is the roof area suitable for drainage into the cistern? These rain-collecting surfaces are called "catchment basins," and though they are generally limited to house and farm building roofs, catchment can also be made from a paved ground surface. With this information you can judge whether rainfall will be sufficient to take care of your water needs.

If the amount of water you receive is less than the amount required by your family, you may have to view a cistern as a supplemental source, but not as a primary water source. If you must use a cistern as your sole source, you can reduce your water consumption, increase the

Table 8-3: Capacities of Various Sizes of Cisterns
(measured in gallons)

Depth (ft)	Diameter of Round Cistern (ft)													
	5	6	7	8	9	10	11	12	13	14	15	16	17	18
5	735	1,055	1,440	1,880	2,380	2,935	3,555	4,230	4,965	5,755	6,610	7,515	8,485	9,510
6	882	1,266	1,728	2,256	2,856	3,522	4,266	5,076	5,958	6,906	7,932	9,018	10,182	11,412
7	1,029	1,477	2,016	2,632	3,332	4,109	4,977	5,922	6,951	8,057	9,254	10,521	11,879	13,314
8	1,176	1,688	2,304	3,008	3,808	4,696	5,688	6,768	7,944	9,208	10,576	12,024	13,576	15,216
9	1,323	1,899	2,592	3,384	4,284	5,283	6,399	7,614	8,937	10,359	11,898	13,527	15,273	17,118
10	1,470	2,110	2,880	3,760	4,760	5,870	7,110	8,460	9,930	11,510	13,220	15,030	16,970	19,020
12	1,764	2,532	3,456	4,512	5,712	7,044	8,532	10,152	11,916	13,812	15,864	18,036	20,364	22,824
14	2,058	2,954	4,032	5,264	6,664	8,218	9,954	11,844	13,902	16,114	18,508	21,042	23,758	26,628
16	2,342	3,376	4,608	6,016	7,616	9,392	11,376	13,536	15,888	18,416	21,152	24,048	27,152	30,432
18	2,646	3,798	5,184	6,768	8,568	10,566	12,798	15,228	17,874	20,718	23,796	27,054	30,546	34,236
20	2,940	4,220	5,760	7,530	9,520	11,740	14,220	16,920	19,860	23,020	26,440	30,060	33,940	38,040

Source: *Private Water Systems*, MWPS-14, Fourth Edition (Ames, IA: Midwest Plan Service, 1979), p. 55.

catchment area by guttering more buildings, or increase the catchment by paving some land and leading the runoff into the cistern.

Next, determine how much water you can expect in the wettest years. This figure will tell you how big the cistern will have to be in order to prevent waste and overflow.

In a wet year with precipitation of 45 inches, the hypothetical 944 square feet of catchment will collect 17,747.2 gallons per year (that is 944 × 18.8 gallons). In theory, then, this family of four requires a cistern 13 feet in diameter, and 18 feet deep.

Springs

The person who finds a spring on his property probably has an acre of four-leaf clovers and always snaps off the biggest part of the wishbone. He's a darn lucky guy!

Springs, you will recall, are caused by groundwater gushing from a crack or fault in the earth. Somehow springs have a mystique that wells do not have. Perhaps people find something magical in water that suddenly spurts from otherwise dry land.

But that mystique leads many to think that all springs, without exception, are clean and pure. Unfortunately, many are not.

Because their source is groundwater, springs are susceptible to all the ills and poisons that seep through the ground and enter an aquifer. Therefore, water springing from an aquifer contaminated by surface water, or sewage, or chemicals will be as polluted as any other groundwater source. A spring's magic does not extend to powers of self-cleaning. That is your job, should you choose to do it.

If you decide to use your spring as a drinking water source, you must know two things: does the spring give an adequate supply of water and is the water pure enough to drink as is, or with a little treatment? Secondary considerations have to do with the location of the spring. Is it uphill of your home so gravity will bring the water to your tap, or will you have to pump the water? Finally, what kind of encasement or storage tank must you build to hold a sufficient supply to meet peak demand?

Begin by measuring the spring's flow. If the supply is insufficient, there's no point in going further. Be sure to measure the flow during

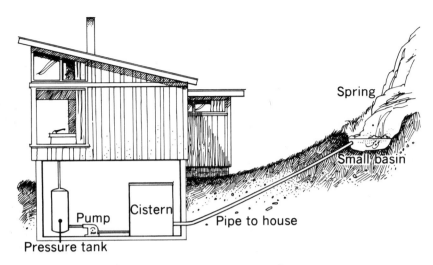

Figure 8-7: Gravity Feeds Spring Waters into Cistern.

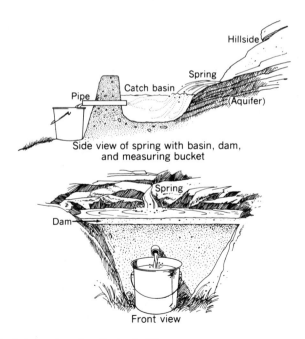

Figure 8-8: Measuring the Spring's Flow.

the driest season for your locale. You will have to dig out a spot just below the spring to form a small basin. With the dirt from that hole, and maybe some additional soil as well, make a dam to hold back the water so the catch basin fills. As you build the dam, place a pipe in the center of the dam, just slightly below the water level in the basin.

Put a measuring bucket just below the dam's water spout. (You may have to level the ground to keep the bucket from tipping.) Then simply clock how many gallons the spring gives forth in a minute, or five-minute period. From that, calculate how many gallons a day the spring will produce for you. Is the amount sufficient? Compute the drinking water needs of each person in your family—about 2 quarts a day for each, plus additional amounts for cooking. If you are very lucky, your spring may supply enough water for all household needs.

If the supply is sufficient, then test the water for coliform contamination, heavy metal pollution, and nitrates. (See Chapters 4 and 5 for testing procedures.)

Once you have established that the flow is adequate and the water clean, all you have to do is catch it and bring it to your home.

Either a catchment basin or a storage tank can be built to hold the water. It should be big enough to hold a half-day supply. It can be constructed of reinforced concrete, and will need a cover. It also must

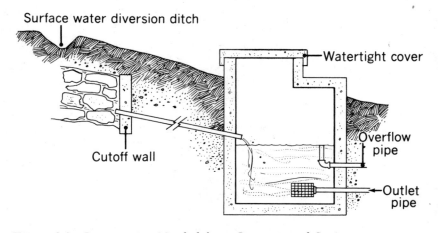

Figure 8-9: Construction Needed for a Concentrated Spring.
(Source: Private Water Systems, *MWPS-14, Fourth Edition [Midwest Plan Service, Ames, IA 50011, 1979], p. 23, with permission of the publisher.)*

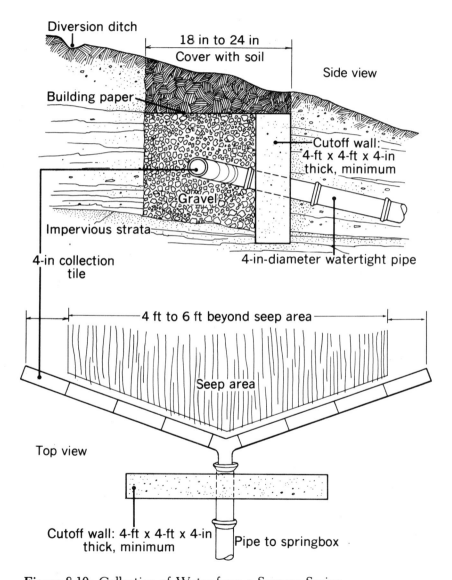

Figure 8-10: Collection of Water from a Seepage Spring.
(Source: Private Water Systems, *MWPS-14, Fourth Edition [Midwest Plan Service, Ames, IA 50011, 1979], p. 25, with permission of the publisher.)*

have an opening so someone can get inside to clean and maintain it. It will need an overflow drain, and a household plumbing hookup.

The storage tank can be located in a basement, or a catchment basin can be built at the spring, and tailor-made to fit the location. If you decide on a catchment basin rather than a storage tank, you must place it deep enough into the earth to ensure sufficient storage capacity. You will also want to dig a drainage ditch above and around the spring to catch surface water and divert it away from the spring. Additionally, you'll have to put up a fence to keep out livestock, wildlife, and even the family pets.

If the spring is located uphill from your home, you can rely on gravity to bring the water to you. However, the elevation must be at least 20 feet. If the spring isn't located high enough, the water can flow slowly into the storage tank or cistern. From there, it can be pumped into the taps with sufficient pressure.

Some springs require additional development. For those that seep rather than spurt, it is necessary to line the boggy area with tile. In that way, the seepage can be caught and made to flow efficiently into a storage area.

Any piping connecting the spring to the house should be laid deeply enough to protect it from freezing—at least 3 to 4 feet in most areas of the country, but deeper in many places.

For additional information about constructing a well, cistern, or spring, write the National Water Well Association, Information Officer, 500 West Wilson Bridge Road, Worthington, OH 43085, or call (614) 846-WELL.

NOTES

1. *America's Priceless Ground Water Resource: Fact and Fiction. The Story of Underground Water* (Worthington, OH: National Water Well Association, 1977).

2. Shingara S. Sandhu, William J. Warren, and Peter Nelson, "Magnitude of Pollution Indicator Organisms in Rural Potable Water," *Applied and Environmental Microbiology*, April, 1979, pp. 744–749.

3. Lauren Brown, *Grasses: An Identification Guide* (Boston: Houghton Mifflin Co., 1979), p. 89.

4. Barbara and Ken Kern, *The Owner-Built Homestead* (New York: Charles Scribner's Sons, 1977), p. 73.

5. Tom Williamson, "Dowsing Achieves New Credence," *New Scientist,* February 8, 1979, pp. 371–373.

6. Agricultural Research Service, Farmers' Bulletin No. 2237, *Water Supply Sources for the Farmstead and Rural Home* (Washington, DC: U.S. Department of Agriculture, 1971), p. 14.

The Future:
9 Crisis, Conservation, and a Silver Lining

Previous chapters of this book have explained how you can take action to help protect your own family from contaminants in drinking water. Hopefully, you now have formulated a plan of attack, and have developed a system of treatment that may help protect the health of everyone in your household.

The benefits of purer water are many and far reaching. But we have long been denied one of the simplest of all pleasures—drinking from an abundant, unpolluted water supply. Since the Middle Ages and probably before, people have been afflicted by terrible waterborne diseases. Today, we have the added worry of chemical contaminants in water. As a result, the task of providing life's basic necessities—food, air, and water—has become a time-consuming and confusing pasttime instead of a simple matter of eating, breathing and drinking those good elements provided by Mother Nature.

But by taking matters into your own hands, rather than relying solely on the judgment and efforts of others, you can attempt to ensure yourself and your family of good water. And you can at least feel somewhat confident that you're in control and that your children are much less likely to have inside them a terrible timebomb set to explode into malignancy in twenty or thirty years—at least no timebomb set by the water they drink. The psychological benefits of knowing you are doing something about your water problems can be enormous.

200

And now that you have worked to improve your water, certainly you will hold it in greater value than ever before. Others may spill and waste it, but you will be closer in spirit to the desert Indians who prized and used each rare drop carefully.

At this point, we might think the job is finished. But really it is not. As conscientious people, we have a lot of work ahead. Water experts and ecologists alike have predicted a shortage of water that will have enormous impact on the future of the United States. Some believe the coming water shortage will make the energy crisis of the seventies look like mere child's play. By acting quickly to adopt water conservation measures, this crisis can be averted.

The amount of water used and wasted in this country far exceeds that of any other nation in the world. By learning to conserve now we can spare ourselves much deprivation in the future.

Conservation at Home

Early in this book, we made the statement that there is a set and unchanging amount of water in the world. There is no more and no less water today than when the earth began. How, then, can there be a water shortage?

While the amount of water is unchanging, its location can change radically. Take, for example, the water shortage that is just developing in New York's Long Island. This island is the largest in the continental United States, extending 118 miles into the Atlantic Ocean. It is separated from New England on its north by the Long Island Sound, and separated from Manhattan on its west by the East River.

An encyclopedia published more than ten years ago will tell you that most of the island's population and industry are found at its western end, in the New York City boroughs of Brooklyn and Queens. At the center of the island, and extending eastward, the encyclopedia states there are more than 3,000 farms covering 153,000 acres, comprising one of the most important truck farming regions in the United States. Other island industries include fishing, oystering, and tourism.

But these facts, true only a few years ago, are no longer true. One field that formerly sprouted those famous Long Island potatoes is now covered by the streets, houses, driveways, and lawns of the community called Levittown. East, west, north, and south of Levittown are more communities just like it, making Long Island a 118-mile-long stretch of suburbia. To be sure, a few farms still exist, the prized ducklings are still raised there, and some deep sea fishing still takes place at Montauk and a few of the eastern-most harbors. But mostly, Long Island has been "developed." And that is the cause of its current water shortage.

When the land was farmed, rain falling to earth was received by the soil. It percolated downward, until it reached the aquifer where it was stored. The relatively few wells on the island tapped an abundant source of water. However, as the island was paved for housing developments, shopping centers, industrial parks, and highways, it also was fitted with a system of storm sewers. As a result, the water that falls to earth runs off roofs and pavements, and is diverted into culverts that carry it to the Atlantic or Long Island Sound.

At the same time this necessary part of the hydrological cycle—the replenishment of the aquifer—was bent out of shape, the water demand of a burgeoning population increased enormously. Individual and municipal wells pumped mighty drafts from an aquifer that received a greatly diminished recharge of water. As a result, the water table lowered dramatically. Salt water from the surrounding ocean now threatens to enter the underground supply. Even as the public is made aware of the threat, a massive sewer project is in the works. It will divert still greater amounts of fresh water to the ocean.

In essence, Long Island is a small example of what is happening all over the country. The Sun belt states have seen massive development in the last two decades. Residents of southern California have felt the pinch of drought along with residents of Las Vegas, Phoenix, and Tucson. In these states, where rainfall is only half (at best) of Long Island's annual 39 inches, the water table will decline rapidly as residents indulge in a kind of "deficit spending" of water.

In an area of the west called the Texas High Plains region, which encompasses parts of Texas, New Mexico, Oklahoma, Colorado, Nebraska, and Kansas, water has been drawn from a great underground body of water called the Ogallala Reservoir. Used primarily for irrigation, it has made the desert bloom. That region now produces about 20 percent of the United States' cotton and sorghum crops, and supports herds of beef cattle. The population of the United States has in fact become dependent on this region for a good amount of its meat.

In the Texas High Plains, the water table has dropped an average of 2 to 7 feet annually, and by the year 2000 all the water in the Ogallala Reservoir, which took millions of years to develop, will be gone.[1]

In parts of Pinal County, Arizona—where Tucson is located—the ground has dropped as much as 12 feet because the water table supporting the ground has fallen drastically.[2] The residents of Tucson, in fact, have used up half of the water in their underground aquifer, a body of fossil water.

Bruce Johnson, a hydrologist for Tucson's water supply, believes the main problem is that rain cannot replenish the aquifer's water as fast as it is drawn out. Nevertheless, huge draughts are drawn each day to keep Tucson's lawns green.

"But," says Johnson, "we are schooling people about ways to conserve water." He notes the total amount used for 1977–78 has dropped back to the same level used in 1972–73. Home conservation will not be the total solution for Tucson, however, and Johnson believes the city may have to dig new wells. Eventually, the city will need water from an ajacent aquifer, and possibly become another mouth sucking at the bosom of the Colorado River.

Other sections of the country are also headed for trouble. The Environmental Protection Agency has predicted that by 1990 water shortages will exist in about 200 New England cities and towns.[3]

If people were forewarned of the impending crisis, certainly they would use water more cautiously. The lower the demand for water, the less it has to be processed, and purified for drinking, and less waste water will be produced. Nevertheless, household conservation requires careful thought because most household appliances such as washing machines and garbage disposal units have not been designed with water conservation in mind. Neither have plumbing fixtures like toilets and showers. As a result, an enormous quantity of water is used for domestic purposes other than drinking and cooking.

The average person uses 70 gallons of water each day inside his home. This figure does not include water used for lawns, gardens, swimming pools, ponds, or car washing, but represents the amount used in drinking, cooking, personal hygiene, and household cleaning tasks. Toilets use 45 percent of indoor residential consumption, or 32 gallons. Bathing, shaving, etc., use 30 percent, or 21 gallons. Laundry and dishes use 20 percent, or 14 gallons, and drinking and cooking use only 5 percent, or 3 gallons.[4]

Since the average toilet uses the major portion of household water, it is first in line for reform. Drought-wise Californians quickly learned to reduce the average 5-gallon flush by placing bricks or filled plastic bottles inside the toilet tank to displace some of that water. Under pressure from conservationists, some manufacturers have designed toilets that combine the water with air, so that an efficient flush can be completed with only 2½ gallons of water. Others have designed units with a dual switch, allowing someone to use 2½ gallons for a regular flush, or 1¼ gallons for flushing only liquid waste. In addition, composting toilets and chemical units have become increasingly accepted by the public.

But by far the greatest water saver in regard to toilets is simply fixing a leaking or "running" toilet. Leakage can be responsible for as much as 10 percent of all residential water consumption. Some experts estimate that the life of a toilet tank valve is seven years, and after that time toilets are prone to leaks. Even a very small leak can consume enormous amounts of water.

Other household appliances can conserve water if you use them with that goal in mind. Fill washing machines and dishwashers to capacity before running them. Unplug your garbage disposal. Think before you turn on the faucet. And, of course, you can try what Californians recommend, and "shower with a friend."

Every drop of water saved from unnecessary use increases the total supply for everyone else.

However, we know that domestic conservation is not the sole answer to the crisis of water quantity or water quality. In fact, most water is not used domestically, but rather by agriculture, industry, and power plants. An international group called the Organization for Economic Cooperation and Development, comprised of twenty-four nations, recently evaluated the quantity and quality of water used by its members. The survey noted that water usage varied greatly among nations—with the average American using ten times the water of the average Briton—and that in all countries water use is increasing.[5] However, this increase is not caused by household use, which is actually on the decline in most countries. Rather, agriculture and power plant production claim more than the lion's share of water, with these demands increasing annually.

In the United States, we have seen agriculture's water requirements grow as the American diet has become more affluent. Whereas

2 pounds of edible grain requires 240 gallons of water to produce,[6] 1 pound of beef requires 4,000 gallons of water.[7] And Americans are eating more beef and fewer grains and vegetables in a misguided effort to achieve a healthy diet.

In fact, today agriculture accounts for 47 percent of the nation's water use; only 9 percent is used domestically. Probably the greatest waste of water results from irrigation, where much is lost to evaporation and to unplanted soil. An alternative method, one that will become cost-effective as water becomes increasingly scarce, is drip irrigation, where water trickles slowly and directly to the roots of growing plants.

Industry, too, hogs more than its fair share of water, about 20 percent: the production of 1 ton of steel requires 30,000 gallons of water, and the manufacture of the paper in this book requires 184,000 gallons.[8] But the greatest amount of water is used in the production of energy. Water is needed in vast quantities to move coal slurry out of the mines, it is needed to remove oil from shale, and great quantities are needed to cool nuclear reactors, or to generate steam for electric power plants. Unfortunately, the area of the country with the greatest resources of coal, shale, and uranium is the still undeveloped Rocky Mountain region—which is as water-poor as it is resource-rich.

Maintaining Water's Quality

The increased demand for water has a direct effect on the quality of water available. While some communities can now turn up their noses at the idea of using polluted river water as a drinking source, water may soon become so scarce that many will be forced to do just that. Once these low-quality water sources are tapped for drinking, more chlorine will be added to combat bacteria and viruses. The final product will be laced with chlorine and halogenated organic chemicals that are dangerous to our health.

What, then, can be done? Some Arab states have toyed with the notion of towing an iceberg to their country to supply good water. In the United States there has been discussion of piping Yukon River water from Alaska to the lower forty-eight—at the cost of billions of dollars. Still others look to the oceans for drinking water. Once, desalinization was impractical because of its high cost. For a long time the only method available was distillation. However, today's reverse osmosis units can remove the salt from sea water at 20 percent below the cost of distillation.[9] Desalinization plants are in operation in Saudi Arabia, Algeria, Russia, Venezuela, and California. Supporters of desalinization see it as the solution to a global water crisis, especially since the oceans hold 97 percent of all the earth's water. However, conservation is cheaper and is something everyone, everywhere, can do right now.

In addition to conserving our water, it is also necessary to safeguard its purity. While federal laws exist today to keep our surface and underground water supplies from being poisoned, they are often ignored or frequently not enforced with vigor. Many treatment plants overflow during periods of wet weather, pouring raw sewage into streams and rivers. Officials shrug and place the blame on the rain and the high cost of improvement or plant expansion. In fact, the biggest immediate threat to the purification of our basic water supply is probably the cost of the programs involved.

As inflation and the economic slump take a greater toil, many industries will ask to be excused from environmental regulations in the name of cost-cutting or inflation fighting. Often the government is willing to go along with proposed "economy measures."

Here's a good example of what can happen in the name of economy. The EPA has proposed regulations under the safe drinking water act for the amount of organic chemicals (chloroform, THMs, and industrial wastes) in our drinking water. But before the regulations were even in force—and they apply to only about sixty cities with a population of more than 75,000—they were criticized by two separate water works associations and one government agency as being unnecessary, too expensive, and inflationary. The National Association of Water Companies, representing privately owned water utilities, said the ruling was premature, and inflationary. The American Water Works Association (AWWA) found the proposed limit on contaminants too stringent, and the EPA's estimated cost of "under $500 million" to install granular activated carbon (GAC) beds to be unrealistically low. The AWWA estimates the cost may reach $5 billion. Eventually, the government's own Council on Wage and Price Stability got into the act, saying EPA's demand for GAC beds was inflationary, suggesting the agency find some cheaper way to treat water to remove organic chemicals. All these criticisms are leveled at an agency that acted to regulate levels of organics only after being sued by the Environmental Defense Fund.[10]

While the cost of GAC is hotly debated in this country, many European cities have been using this procedure since the early sixties. They have developed the hardware and procedures to recycle the carbon on site, making the treatment economical as well as effective in removing trace organics.

According to Gordon Robeck, the director of drinking water research at the USEPA, the use of GAC beds has not been an inflationary factor in water treatment in Europe, and doesn't necessarily have to be inflationary in this country either. "Of the overall cost of delivering water, only about 16 percent is due to the cost of treatment. The remainder is generated by the expense of equipment, water mains, labor, management and maintenance. The smaller the city, the greater the cost of treatment, proportionally."

Without the vigilance of environmentalists and the enduring concern and education of the American public, it seems little would be volunteered by either industry or government to improve our environment and health.

But the future is not completely bleak. Even while gypsy haulers are illegally dumping battery acid into our rivers, and chemical companies are planting their wastes in our soil, small signs of scientific gain are surfacing around the world. Following is a brief sketch of what scientists are finding. Significantly, these developments represent true steps forward in water treatment. Each involves the use of a natural product. None involves any manufactured chemical with an unknown effect.

In Southeast Asia a company located in Bangkok, Thailand, has developed a system of water filtration for rural areas.[11] The two-stage filter uses fibers from shredded coconut husks to filter suspended solids out of water and then "polishes" the water by passing it through a filter of burnt rice husks to remove any residual turbidity.

Using very cheap, very available materials, and *no* chemicals, the filter significantly improves the taste, odor, and color of drinking water. It removes enough solid material from the water to meet the World Health Organization's international drinking water standards for particulate matter in water. While it removes only 60 to 85 percent of the coliform bacteria in water, and therefore cannot be depended on for bacteriological improvement, the filter significantly and cheaply improves water that otherwise would be left completely untreated. It represents a simple first stage investment in what will become a multi-stage purification process.

In Göppingen, West Germany, the city's water supply is monitored for the presence of heavy metals by fish. Yes, fish! The water utility imported six very special fish from the Nile. Called elephant fish, they have the ability to emit several hundred small electrical impulses per second. They are extremely sensitive to cadmium, zinc, copper, mercury, and other heavy metals. When these toxins are present in the water supply, the fish give out fewer and fewer electrical signals.

The water utility in Göppingen has designed a foot-long aquarium rigged with an alarm. If the fish's electrical impulses drop below 200 per second, the alarm sounds and the water is tested at the utility's lab.

Each of the fish spends a lonely two-week vigil in the aquarium, monitoring the quality of drinking water day and night. After its shift is completed, the finned guardian joins the others in a larger tank maintained by the company.

These fish are probably the cheapest water utility employees in the world, and have so far been reliable in monitoring for heavy metals. The experiment has been so successful and economical that the town of Ulm, 20 miles south of Göppingen, plans to copy the system.

In Bay St. Louis, Mississippi, biochemist B. C. Wolverton, Ph.D., a senior research scientist for NASA, determined that a weed called the water hyacinth has the ability to purify water.[12]

The weed is highly prolific, and much hated in the south, where it clogs canals and waterways. The state of Louisiana has spent millions trying to control its growth and proliferation.

But Dr. Wolverton found this pesky weed absorbs pollutants from water. The roots naturally absorb lead, mercury, strontium-90, plus organic chemicals—concentrating these substances as much as 10,000 times from the surrounding water.

Dr. Wolverton says that the plants used for domestic waste treatment (as opposed to chemical waste treatment) can be eventually fermented to produce methane, then dried, and cleansed to make compost fertilizer, animal feed, mulch, and even for a protein source.

"I fully intend to solve a major pollution problem, a major energy problem, a major food problem and a major fertilizer problem," says Dr. Wolverton.

In Saskatchewan, Canada, the Engineering Division of the Saskatchewan Research Council heard of the water hyacinth's purification ability and set about finding whether any plants native to Canada, with its short growing season and very cold winters, could duplicate it.[13]

Sure enough, after laboratory tests on local cattails and bulrushes, the Canadians concluded their hardy northern plants had as much potential for water purification as any southern hyacinth.

The research council, in a series of experiments conducted during 1976 and 1977, grew common cattails and bulrushes in laboratory trays, and fed them raw municipal sewage. The plants were initially monitored only for their ability to remove total phosphorus and total Kjeldahl nitrogen from the sewage. High rates of purification—up to 98 percent—were achieved in less than three week's growing time.

The experiment report states that, during twenty experimental runs, the plants showed an "unabated ability" to remove nutrients from sewage. Moreover, the rates of uptake by the plants exceeded the rates predicted.

The research group is sure these plants can be grown to treat municipal waste waters for both rural and urban communities. They propose digging shallow growing lagoons where the plants can do their work, with dirty water scheduled to enter, remain for purification, and be discharged as clean water according to the plants' growing season.

The report points up a second important potential application: the use of these plants to treat runoff from barnyards and livestock feed lots. Employment of plants in these areas would greatly ease pollution of both ground and surface water.

Moreover, researchers say the aquatic plants, after removing the nutrients from the runoff, could be harvested for animal feed. They found that plants grown in this sort of nutrient rich environment have a higher crude protein content, and more digestible organic matter. The report states, "In a feedlot environment the technique offers a complete system, from nutrient removal to recycling the nutrient-rich plants as animal feed, soil conditioners, or bedding material in the corrals. This should make economic sense in view of the rising cost of energy and animal feed."

Preliminary studies have shown these plants also can effectively remove mercury, phenols, and PCBs from water and sediment.

The economic benefits of developing native aquatic plants to purify water are very great. The health benefits of using effective natural agents—with no chemicals added—can be astounding.

In the Atlantic Ocean, below the Continental Shelf, scientists looking for oil found fresh water instead.[14]

A United States Geological Survey gathering geological samples on the ocean floor discovered a zone of fresh water. It extends from the Georgia coast to Georges Bank off New England, 60 miles seaward at the East Coast, and beneath the Continental Shelf.

Chief scientist for the expedition, John C. Hathaway, found the water is significantly less salty than sea water, and "will be useful in planning development of water resources in coastal areas." He believes most of the water entered the sediments of the Continental Shelf during the Ice Age when the sea was much lower than it is today.

Perhaps Long Island will find its future water supply somewhere off Jones Beach.

And, finally, in Cambridge, Massachusetts, scientists at the Massachusetts Institute of Technology (MIT), with the support of a federal grant, are looking into the marvelous properties of chitin (pronounced KITE-in).[15]

Chitin is found in the hard covers of shellfish like crabs, lobster, and shrimp, a commodity that was considered a major nuisance to the fish processing industry. Ever since federal regulations prohibited the dumping of shells into the ocean, the industry has been overwhelmed with shell castoffs. But now scientists say that chitin derivatives can purify water, heal wounds, and even be manufactured into food wrap.

The world's first chitin conference was held at MIT in mid-April of 1977. Scientists there said chitin can be processed into chitosan, a white powder that can purify water so efficiently it may soon outclass all other methods. It can even pick up heavy metals, suggesting it as a possible agent to remove even radioactive wastes from water discharged by nuclear plants.

Ray Pariser, associate director of MIT's Sea Grant Program, said, "With a quarter of a pound of chitosan, you could purify many thousands of gallons of water. . . . You could produce chitosan for between one and two bucks a pound and hopefully much less in the future." He notes that a 5-pound lobster yields about a quarter pound of chitosan, which is biodegradable.

While interest in chitosan is relatively recent in the United States, Pariser points out that Japanese firms have been using chitosan for water purification, and he believes "they're making a good profit."

Perhaps the future is to be a rosy one. Picture the sun rising over lagoons of waving cattails and bulrushes, in a land where fishermen wade into crystal streams to snare fat, healthy trout. Imagine harbors where barge upon barge of waste shells are waiting to be processed into a cheap and effective filtration material for our cities' reservoirs. Imagine rural homesteads where basement tanks grow luxuriant water hyacinths that remove stray toxins from the well water. And everywhere, elephant fish in tiny aquariums keep eternal watch over the purity of our supply.

Does the description sound too far out? Too good to be true? Perhaps if we break free of the philosophy that man can invent chemicals that solve our problems, and apply the same amount of intelligence, money, and creativity to utilizing the products of nature, our future health and environment will be better than anyone can imagine.

Notes

1. Nicole Ball, "Water: The Real Constraint on American Food Production," *The Ecologist Quarterly*, Spring, 1978, pp. 20–31.

2. Michele Strutin, "Pulling the Plug on Arizona," *Mother Jones,* August, 1979, pp. 16–21.

3. "The Conservation Connection," in *Waterways,* January, 1979, p. 1.

4. Murray Milne, *Residential Water Conservation* (Davis, CA: University of California, March, 1976), p. 19.

5. "The Global Water Environment," *Environmental Science & Technology,* July, 1979, pp. 786–787.

6. Lester R. Brown, *The Twenty-Ninth Day: Accommodating Human Needs and Numbers to the Earth's Resources* (New York: W. W. Norton and Company, 1978), p. 142.

7. Kathleen K. Wiegner, "The Water Crisis: It's Almost Here," *Forbes,* August 20, 1979, p. 62.

8. Ibid.

9. "Getting Drinking Water from the Sea Is Becoming Economic," *The Economist,* August 4, 1979, p. 67.

10. Steven J. Marcus, "Troubled Waters," *Technology Review,* February, 1979, pp. 14–15.

11. Richard J. Frankel, "Operation of the Coconut Fiber/Burnt Rice Husks Filter for Supplying Drinking Water to Rural Communities in Southeast Asia," *American Journal of Public Health,* January, 1979, pp. 75–76.

12. "Water Hyacinth: Weed with a Big Future," *Organic Gardening and Farming,* November, 1976, p. 103. Personal communication with B. C. Wolverton, Ph.D., Senior Research Scientist, NASA, November, 1979.

13. G. Lakshman, "An Ecosystem Approach to the Treatment of Waste Waters," *Journal of Environmental Quality,* July–September, 1979, pp. 353–361.

14. Personal communication with John C. Hathaway, Chief Scientist, U.S. Geological Survey, November, 1979.

15. Personal communication with E. R. Pariser, Associate Director of MIT's Sea Grant Program, November, 1979.

Appendix A: A Glossary— All about Water—From H_2O*

activated carbon — a substance used to purify water. The "active" type of carbon is formulated mainly from coal and lignite. It is used in public water filtration systems and some household water filters. Running water through activated carbon eliminates odors, off tastes, and some impurities. Granular activated carbon, in a good filter, will also remove chlorine and organic chemicals.

aeration — a water treatment procedure in which water is combined with air by spraying, bubbling, or trickling. The method removes odors produced by algae, oxidizes iron and manganese, and improves the flat taste of cistern or distilled water. Most gasses and dissolved compounds that easily evaporate or vaporize will be removed by aeration. These include hydrogen sulfide, many hydrocarbons, chloroform, and other chlorine compounds. The rate of removal will depend on temperature. At lower temperatures, many hydrocarbons, like chloroform, will be lost slowly.

alum — a chemical substance (aluminum ammonium sulfate) used by public water treatment plants to purify water. The use of alum can significantly increase the aluminum concentration of drinking water.

Prepared by Jean Polak, Editorial Coordinator, Rodale Press

213

American Bottled Water Association — issues purity standards for commercially sold bottled water. The agency helped bring about federal regulations for bottled water by coordinating rules that previously varied from state to state. For more information, write the American Bottled Water Association, 1411 West Olympic Boulevard, Los Angeles, CA 90015, or call (213) 384-3177.

American Water Works Association — a nonprofit, scientific, and educational organization of more than 31,000 people interested in drinking water supply. It issues publications, develops standards, and supports research programs in water works design, construction, operation, and management. American Water Works Association, 6666 West Quincy Avenue, Denver, CO 80235, or call (303) 794-7771.

aquifer — a porous rock formation that holds water. The formation can consist of sand and gravel or a layer of sandstone or cavernous limestone. At the bottom of an aquifer is a layer of bedrock which makes the formation fairly water-tight. When a well is constructed, it taps the water contained in an aquifer.

artesian wells — mineral waters naturally forced to the earth's surface by internal pressure. One of the oldest and best known artesian wells is Mountain Valley Water located in Hot Springs, Arkansas.

asbestos — a mineral that separates easily into long, flexible fibers that resist fire and most solvents. Prolonged exposure to asbestos fibers can result in pulmonary fibrosis (asbestosis), emphysema, and lung cancer. Although asbestos naturally occurs in certain water supplies in small concentrations, massive amounts of the mineral are being found in the drinking water of large American cities. It appears that the asbestos cement pipes used to carry the water are disintegrating, thereby releasing asbestos into the water.

backwashing — a term used to describe the reversible flow of water (bottom to top) through the resin tank of water softeners, or in filters, to dislodge the accumulated sediment and mineral particles in the exchange bed or filtration medium.

bacterial contamination — a category of microbiological contaminants consisting of bacteria, viruses, and parasites that can cause waterborne diseases. Public drinking water supplies are tested daily for bacterial contamination. Chlorination is the usual disinfection treatment.

bottled water — potable water sold commercially. Some is mineral water, some spring water. Other bottled waters may be purified tap water. The four categories of bottled water include: drinking water, distilled drinking water, fluoridated water, and purified water.

bromine — a chemical germ killer similar to chlorine used mainly to disinfect swimming pools.

calcium — a mineral responsible, in part, for hard water. Studies have shown that the dissolved calcium in hard water limits the internal absorption of the heavy metals lead, cadmium, and zinc.

carcinogens — descriptive name given to substances that cause cancer.

cascade aerators — water treatment equipment patterned after the babbling brook principle. Water is spread out so that it is shallow, then forced over many obstructions to produce turbulence. That allows the water to come in contact with air, which is the cleaning medium. Water is oxygenated and degassed of those gasses having partial pressures below those present in air.

catchment basin — rain-collecting surfaces on house or farm building roofs suitable for water drainage into a cistern.

chemical solution feeders — automatic pumps used to add treatment chemicals to water at a constant ratio. For the homeowner who has several water problems to treat, a chemical feeder is an essential component of his water system. Feeder pumps are usually wired to operate with well pumps, and work by adding the chemical solution into the water line between the well pump and pressure tank.

chlorine — or chlorinated lime is the chemical used to disinfect water.

Since its use began in 1908, chlorine has been used to kill the bacteria, parasites, and some viruses that cause waterborne diseases and to improve the taste and color of water. Most recently, studies have shown chlorine, when used to disinfect water, interacts with humic acid to form chloroform, a known cancer-causing agent.

chloroform — a chemical once used as an anesthetic, but banned by the Food and Drug Administration in 1976 because of its carcinogenic properties. Chloroform can be found in drinking water as a result of the chlorination process.

chloro organics — a group of organic chemicals often found in the drinking water. The most common are compounds known as trihalomethanes (THMs), of which chloroform is the most prevalent. Others include vinyl chloride, and carbon tetrachloride. All are suspected cancer causers. The Environmental Protection Agency has set 0.10 milligram per liter as the maximum allowable contaminant level for THMs.

cistern — a tank, usually underground, for storing rainwater. This water can be used for drinking or it can be a practical source of additional water for laundering and hot water heating.

coliform bacteria — in drinking water, indicates that human or animal wastes are entering the water supply, and other harmful contaminants may also be getting in. Coliform bacteria, by themselves, are not harmful since they are found in the intestines of humans. The coliform bacterial count in the drinking water acts as a gauge or measuring stick for other forms of bacterial contamination.

corrosion — a natural process in which acid water gradually eats away the metals found in household plumbing. As a result, lead, copper, cadmium, and other metals that have an adverse effect on health enter the drinking water system. Other factors in water that can induce corrosion are oxygen concentrations, electrical conductivity, and temperature.

corrosion control — water that is highly acidic is corrosive. Acid water can be identified by measuring pH. The lower the pH number under

7, the more acid the water. Corrosion can be eliminated or reduced by raising the water pH with potassium carbonate or potassium hydroxide. Corrosion problems can also occur if water is highly alkaline. In this case, running the water through a water softener or an ion-exchange medium followed by reverse osmosis treatment will solve the problem.

deionization units — water treatment equipment that works on the principle of substituting one ion for another in water. Water softeners are deionization units, with sodium ions substituting for calcium and magnesium ions in water. Other deionization units use hydrogen to replace unwanted element ions in water. These units have two ion exchange materials—one to remove positively charged ions and another to remove negatively charged ions. Water treated by deionization will contain few, if any, minerals. Deionization will remove copper, fluoride, hardness, soluble iron, nitrates, silica and silicates, sodium, sulfate, total solids, arsenic, and selenium. It will not remove bacteria or organic chemicals.

disinfection — the destruction or killing of disease-causing bacteria. Water can be disinfected by chlorine, ozone, iodine, bromine, silver, and ultraviolet light.

distillation — a process whereby water is changed to vapor through heating and then is allowed to condense back into liquid form. Distilled water contains no solids, minerals, or trace elements, and distillation is one method the homeowner can use to remove contaminants from the drinking water.

downspout filters — a water filter placed in the downspouting to clean the rainwater before it enters the cistern. These filters need to be maintained regularly or they will create more pollution than they remove.

dowsing — the ancient technique of locating underground water using a divining rod. Today the forked willow branch has been replaced by a fiberglass rod or a slim piece of copper tubing. As the dowser walks with his rod held in front of him, the presence of underground water may be indicated when the rod dips downward. The dowsing phenomenon is believed to result from small magnetic field variations.

effervescence — a term used to describe the natural bubbliness of certain bottled drinking water. The bubbliness is produced by the escaping carbon dioxide gas contained in the water. Only naturally occurring bubbling water can be sold as effervescent; bottled waters pepped up with carbon dioxide gas after leaving the spring do not fulfill the effervescent claim.

effluent — the discharge resulting from waste water treatment plants. The effluent usually contains organic human matter, some trace elements, and organic and inorganic compounds from industrial sewers. Effluents are major point sources of pollution.

endotoxin — a substance given off by dying bacteria and known to cause fever when injected into a person's blood stream. Endotoxin can remain in drinking water after normal treatment. The health effects of endotoxin residue in water are not yet known.

Environmental Protection Agency (EPA) — the federal agency, created in 1970, responsible to monitor pollutants in our environment and to set acceptable limits, to develop methods to clean the environment, and to enforce the laws and standards it has developed. Environmental Protection Agency, 401 M Street, S.W., Washington, DC 20460, or call (202) 755-0707.

filter cake — the scum that accumulates inside a filter during water treatment. When filter cake is 1 or 2 inches thick, it has to be removed along with the top inch of the filter medium.

fossil water — underground water trapped millions of years ago and completely encased in a nonporous rock aquifer. Fossil water replenishes itself very slowly, so it could be completely used up before it's replaced. Tucson, Arizona, is an example of a city using fossil water as its drinking water source.

fractional distillation units — water treatment equipment that will remove chloroform and organic chemicals as well as debris, bacteria, minerals, and other contaminants. The basic components of the distiller are the heater, boiling chamber, vapor rising column, and condensing chamber. Distillers operate by heating water until it turns to steam, then condensing the steam back into water. Distilled water is pure, but has no taste.

granular activated carbon filters — water filters containing activated carbon (a form of charcoal treated by high temperatures and steam in the absence of oxygen) as its medium. The filters are very effective in removing chloroform, chlorine, some pesticides, and other organic chemicals, as well as bad taste and odor. They can reduce the concentration of heavy metals, but will not remove fluoride, nitrates, or asbestos fibers. These filters work by adsorption, whereby the surface area of the activated carbon grabs and holds organics and other chemicals. Because bacteria and other organisms grow easily on the carbon surface, the filters must be monitored closely for signs of filter failure and the carbon should be replaced every three weeks or after filtering 20 gallons of water.

groundwater — water that collects below the soil level and is contained by the rock formation or aquifer underlying the surface. Rain is the source of all groundwater. Most of this water is tapped by wells.

hard/soft water — these terms refer to the amount of dissolved minerals contained in water. Hard water has large amounts of minerals in it, while soft water has very little. Calcium and magnesium are the two elements largely responsible for making water hard. Water hardness is measured according to how many grains of dissolved minerals are found in 1 gallon: Water is slightly hard if it contains 1 to 3 grains per gallon; moderately hard if it contains 3 to 6 grains per gallon; hard when it contains 6 to 12 grains per gallon; very hard when it contains over 12 grains per gallon. Studies indicate that drinking hard water may be beneficial to your health, but additional research is needed in this area.

heavy metals — chemically defined as those metals having a high specific gravity or density. Some heavy metals include cadmium, lead, and mercury. A special characteristic of heavy metal chemicals is their strong attraction to biological tissues and, in general, the slow elimination of these chemicals from biological systems. Heavy metals are persistent; once in the system, they remain for relatively long periods of time, continuously inducing their toxic effect.

Hippocrates' sleeve — an ancient form of water treatment recommended by Hippocrates, the father of medicine, in which a cloth bag was used to strain impurities from water.

hydrogen sulfide — a gas that can form in well water and produce a "rotten egg" odor. Certain strains of bacteria feed on the sulfate (sulfur) found in the water and produce the gas as a metabolic by-product. Hydrogen sulfide is very acidic and can cause the water to corrode iron pipes, filters, water tanks, and even concrete. In very high concentrations, the gas is flammable and poisonous; it can cause nausea, illness and, in extreme cases, death. The maximum contaminant level for this gas is 0.05 milligram per liter.

hydrological cycle — or water cycle. A sequence of phenomena showing how water moves in a never-ending revolution. Water in the oceans continually evaporates, putting water vapor into the atmosphere. Precipitation (rain, snow) puts the water on the ground where it can become surface water (lakes, ponds, puddles), or surface runoff (streams and rivers), or seep into the ground and be stored as groundwater. Eventually the water will travel back to the ocean, to begin the cycle again. The exchange between the earth and atmosphere is accomplished by the heat of the sun, the winds, and the pull of gravity.

inorganic solutes — a category of water "contaminants" comprised of the following minerals: barium, cadmium, chromium, lead, mercury, silver, sodium, beryllium, cobalt, copper, magnesium, manganese, molybdenum, nickel, potassium, tin, vanadium, and zinc. While many of these minerals can be quite healthful in small amounts, others can be very toxic in small amounts. Furthermore, some substances that are beneficial in small amounts can be dangerous in large amounts. To further complicate matters, some adverse effects are caused by a combined total intake from water, food, and air.

iodine — a chemical germicide effective in killing bacteria, viruses, cysts, and other contaminants in water. Iodine can be used as a short-term means of water disinfection, but it is not recommended for use by pregnant women, and not considered safe for long-term use.

ion exchange — a common chemical reaction where one ion is substituted for another in solution. Water softeners work on the principle of ion exchange, as do deionization units. The salt nuggets or crystals the homeowner regularly adds to his water softener consist, in part, of sodium ions. When water is softened, it passes through an ion-exchange material—usually salt—where the calcium and magnesium ions are removed and replaced by sodium ions. The softener will also remove poisonous heavy metals like lead and cadmium and can remove most of any radioactivity present.

leachate — the contaminant that forms beneath municipal solid waste landfills when there is enough rainwater left over after evaporation to carry dissolved wastes into the ground, eventually reaching the underground water beneath the landfill. Leachates usually contain measurable amounts of trace metals and organic acids. Water contaminated by landfill leachate usually has a foul odor.

magnesium — the mineral, along with calcium, responsible for making water hard. Studies suggest that the large amounts of magnesium found in hard water can lower death rates from heart disease. However, water with magnesium concentrations of 700 milligrams per liter or more can act as a laxative. Presently, there is no limit for magnesium concentration in drinking water.

measuring elements in water — water hardness is measured in grains. A grain is defined as the average weight of a single, dry grain of wheat which is 1/7,000 of a pound. Very hard water containing 12 grains of elements per gallon weighs 12/7,000 of a pound. Other elements and impurities in drinking water are measured in terms derived from the metric system. Parts per million (ppm) and parts per billion (ppb) describe ratio measurements. For example, 100 parts per million (ppm) of salt in water means that there are 100 parts of salt for every million parts of water. One part per million (1 ppm) is equal to 1 milligram per liter (1 mg/1) and 1 part per billion is equal to 1 microgram per liter (μg/l). A liter is equal to 1.05 quarts, a milligram is one-thousandth (1/1,000) of a gram, and a microgram is one-millionth (1/1,000,000) of a gram.

membrane filter test — one of two methods used to determine the number of bacteria (coliform) contaminating the water. In using the membrane filter technique, a positive coliform test is obtained if more than one coliform bacteria is filtered from 100 milliliters of water. (See also multiple tube fermentation technique.)

mineral water — a term to describe the natural occurrence of large amounts of elements in ground or well water. Mineralization of groundwater can occur from the slow erosion of rock in the aquifer and can cause the water to taste bitter. Highly mineralized water is usually not suitable for drinking. Natural mineralization increases the depth of the well, and may eventually reach a point where the water is no longer drinkable.

multimedia filters — water filters that combine several types of materials to remove contaminants and pollutants. Conventional multimedia filters have the largest particle medium (like gravel) at the bottom, working upward to the top layer, which is made of the finest particle material (like silica sand). In that way, the large particles support the finer ones, keeping them from slipping down the drain. Inverted multimedia filters have the largest particle medium at the top, working downward to the smallest particles. This arrangement allows the larger coarse material to be filtered out of the water before reaching the smaller media particles. The use of inverted multimedia filters provides "in-depth" filtration, where the filtration process takes place throughout the entire depth of the column. That is in contrast to conventional multimedia filters where the top few inches have to filter out the entire contents of the water.

multiple tube fermentation technique — one of two methods used to measure bacterial contamination (cf. membrane filter test) of water. This procedure samples a small volume of water (usually 0.1 milliliter) and reports the coliform bacteria count as the estimated number (most probable number, MPN) of coliforms in 100 milliliters of water. The test is based on gas production (fermentation) resulting from bacterial growth in several test tubes. If the estimated coliform count exceeds 1 bacterium per 100 milliliters, the water is considered unsuitable for drinking.

National Association of Water Companies — an organization representing 225 privately owned water companies. The association conducts research and keeps members informed in economic, legal, and regulatory developments. To contact, write National Association of Water Companies, 1019 19th Street, N.W., Suite 1110, Washington, DC 20036, or call (202) 638-3461.

natural water filtration systems — are being developed by scientists from around the world. Because each of the systems involves the use of a naturally occurring substance or product, they represent actual steps forward in water treatment. Some of these include:

> *Aquatic plants* are being investigated because some have been discovered to help purify the waters in which they live. Water hyacinths, a much-despised water-weed that clogs canals and waterways, has the ability to absorb heavy metal and organic chemicals from the water. Canadian scientists are using cattails and bulrushes to treat municipal waste waters.

> *Chitin* is a substance found in the hard covers of shellfish. Scientists have found a way to process chitin into chitosan, a white powder that effectively purifies water. It is even capable of removing radioactive wastes from water discharged by nuclear plants.

> *Elephant fish* are being used in West Germany to monitor heavy metal concentrations in drinking water. The fish, which have the ability to emit small electrical impulses, give out fewer of these signals when the water contains heavy metals.

> *Filters* made of fibers from shredded coconut husks are being used in Third World countries to remove suspended solids from the water. The water is then passed through a

filter made of burnt rice husks in order to remove any residual turbidity.

nitrates/nitrites/nitrosamines — a related trio of organic compounds known to cause infant illness (methemoglobinemia) and suspected of causing cancer. In water, the major sources of nitrates are septic systems, animal feed lots, agricultural fertilizers and manured fields, industrial waste waters, sanitary landfills, and garbage dumps. Once inside the body, the relatively harmless nitrates are changed into nitrites. In adults, the nitrites can combine with other substances in the stomach to become nitrosamines. By distilling their drinking water supply, homeowners can remove nitrates.

nonpoint source pollution — a contaminant that comes from many different nonspecific sources. For example, both fertilizer and pesticide runoffs from farmlands, contaminating streams or wells, would be considered nonpoint source pollution.

nonpurgable total organic carbon (NPTOC) — a term devised by the Environmental Protection Agency when it tested the efficiency of granular activated carbon water filters. NPTOC is a substance that represents the bulk of organic compounds found in drinking water. These compounds are mostly natural materials that have a high molecular weight.

organic compounds — are natural or synthetic substances based on carbon. Organic compounds can become part of the water supply through water treatment methods, from industry, sewage treatment plants, runoff, and from spills and accidents. Some organic pollutants include: chloro organics (chloroform, carbon tetrachloride), polychlorinated biphenyls (PCBs), polynuclear aromatic hydrocarbons (formed from combustion using fossil fuel and includes fuel mixtures, crankcase oil), herbicides, and pesticides.

Organization for Economic Cooperation and Development — an international organization comprised of twenty-four countries—nineteen west European governments, Canada, the United States, Japan, Australia, and New Zealand. Its purpose is to help stimulate economic cooperation between member countries for economic growth, expand world trade, and coordinate aid to less developed areas in an effort to achieve world prosperity and to contribute to peaceful and harmonious relations among the peoples of the world. Contact by writing the Organization for Economic Cooperation and Development, Publication and Information Center, 1750 Pennsylvania Avenue, N.W., Washington, DC 20006, or call (202) 724-1857.

oxidizing filters — offer a convenient—but limited—one-step solution to iron problems. Oxidizing filters work by converting dissolved iron into insoluble iron by passing the iron-heavy water through a man-ganese-treated material. As the iron forms rust particles, it is filtered out by a granular material in the filter. However, these filters do not work on iron bacteria that can clog the filter and mineral beds. Oxidiz-ing filters must be backwashed and rinsed every week to remove precipi-tated iron from the filter bed.

ozone — a form of oxygen used widely throughout Europe to disinfect water. This unstable gas is an excellent germicide, but it is expensive, it must be produced electrically as needed, and it cannot be stored.

pH — this symbol represents the degree of acidity or alkalinity of compounds. pH measurement is expressed with a number from 1 to 14; 7 is the neutral point. Values from 7 to 1 are increasingly acid, while those from 7 to 14 are increasingly alkaline. The most desirable range for water pH in the home is from 8 to 8.5.

point source pollution — a contaminant that can be traced to an individual source. For example, a factory dumping chemicals by pipe directly into a river.

polyphosphates — a generic term designating food-grade phosphate chemicals used to treat iron problems. Polyphosphates are sold under the names Zeotone, Micromet, or Nalco M-1. When added to the water, they stabilize and disperse the iron so it cannot oxidize when exposed to air. This treatment works only for dissolved iron and is effective in a pH range of 5.0 to 8.0.

precipitation and filteration — two water treatment procedures that combine to produce the most thorough answer to iron problems since both dissolved and oxidized iron are removed. Chlorine, potassium permanganate, or carbon dioxide is added to the water to oxidize the iron. The water is then stored in retention basins so the rust particles can drop to the bottom. Water is then drawn off the top and filtered through activated carbon filters to remove all insoluble iron.

regeneration — the process of cleaning deionization equipment. Regeneration involves the removal of contaminant ions from the resin bed and replacing them with the type of ions originally on the resin bed. The procedure involves flooding the resin with an extremely concentrated solution of the original ion.

residual chlorine — a term applied to the additional amount of chlorine public water systems place in water to kill any bacteria that may be picked up as the water passes through the mains and service lines to homes.

resins — the ion-exchange materials found in deionization units and water softening units. Examples of ion-exchange materials are zeolite and resin beds.

reverse osmosis — a method of treating water that removes turbidity, particulate and colloidal matter, ionized and nonionized dissolved solids, bacteria, viruses, pyrogens, aromatic hydrocarbons, most pesticides, and asbestos. It will not remove simple compounds like chloroform and phenol. Reverse osmosis is a type of pressurized filtration system. The water to be treated is made to flow over a semipermeable membrane that is similar in appearance to cellophane. Under pressure, the water is forced through the membrane, leaving behind the impurities which are unable to seep through. Reverse osmosis units consist basically of the membrane and some filters to remove larger particles of incoming water. No electricity or other energy source is needed to operate the unit.

roof washers — a tank, barrel, or garbage can that is connected to downspouting to catch the dirty water from the first part of the rain. Once the container has filled, water will overflow into a pipe that is connected to the cistern.

sand filters — water filters that use sand as the medium for cleaning the water. Sand filters can remove clay, silt, colloids, and microorganisms. If the pores in the filter are small enough, bacteria and viruses can be strained out. Most sand filters are used as an initial treatment for pond water.

sediment — substances like dirt in water that settle or filter to the bottom.

softeners — a device used in the home to artificially soften water. Most home water softeners work on a principle called ion exchange, whereby calcium and magnesium ions in water are removed and replaced by sodium ions. Fully automatic softening equipment has a timing mechanism that starts and controls the various steps of the softening process. All the homeowner has to do is add salt about every thirty days.

Soil and Health Foundation — a nonprofit organization which provides funds for research and disseminates information to help people grow more of their own food, to better their diets, to upgrade their drinking water, and to improve their health. For a nominal fee of $25, the foundation can test your water for trace element and heavy metal content. For a sample bottle and instructions, write to the Soil and Health Foundation, 33 East Minor Street, Emmaus, PA 18049, or call (215) 967-5171.

spray aerators — water treatment equipment that works by shooting the water high into the air, creating a fountain effect. Spray aerators are not practical on a home scale and cannot be operated during the winter in cold climates.

springs — groundwater that flows to the earth's surface through a crack or fault in the earth.

titration process — a method used by laboratories to measure the amount or concentration of a substance in solution. Water hardness testing kits for homeowner use employ a simplified version of the titration process to count the grains of hardness per gallon.

trace elements — naturally occurring metallic substances found in minute quantities. Our knowledge about the human need for trace elements is sparse because, until the beginning of the 1970s, scientists did little research in this field. Current information indicates that the elements chromium, cobalt, copper, magnesium, molybdenum, selenium, tin, and zinc are essential to human nutrition in small amounts. Barium, silver, beryllium, manganese, nickel, and vanadium are believed to be nonessential, but future research could change this outlook. Cadmium, lead, and mercury are harmful. Most of the trace elements we consume come from food, but the metals can be present in water in a form more readily absorbed by man. Metals in water can come from many sources: the soil, industrial waste, municipal sewage and landfills, the water distribution system, and household plumbing.

trihalomethanes (THMs) — a group of organic compounds produced in water as a result of the chlorination process. These substances form when chlorine interacts with natural humic acid or with algae. Chloroform is the most well known compound in this group. All THMs are known to cause cancer in laboratory animals, and are suspected of being cancer-causing agents in humans. (See also chloro organics.)

turbidity — a term used to describe cloudy water. Undissolved materials such as clay, silt, sand, organic and inorganic chemicals, plankton, and microorganisms combine to give water its cloudy appearance. Turbidity sometimes indicates the presence of sewage, industrial waste, or asbestos. Turbid water can provide an optimum medium for harmful bacteria; therefore, turbidity in water must be reduced to nearly zero for chlorine to be totally effective in destroying germs. Turbidity can be removed by filtration or coagulation.

ultraviolet disinfection — bacteria and certain viruses can be killed by exposing the water to a quartz–mercury vapor lamp that emits ultraviolet rays of the proper wavelength. A home ultraviolet unit consists of one or more germicidal lamps sealed inside a steel cylinder. The cylinder is connected to the water supply line so the untreated water goes into one end of the cylinder, passes over the UV lamp, and then goes out the opposite end into the water line. Ultraviolet light does not work effectively with cloudy or dirty water or with water having a high iron content. Moreover, bacterial spores and some viruses are fairly resistant to ultraviolet treatment. Both the UV equipment and energy needed to run it are expensive in comparison with other treatment methods.

water filters — water treatment equipment that, depending on the type of filter, can remove some turbidity, odor, taste, bacteria, corrosion, chloroform, chlorine, some pesticides, and other organic chemicals like PCBs and PBBs (polychlorinated and polybrominated biphenyls, respectively).

water hardness — a term to describe water containing large amounts of calcium and magnesium. (See also calcium; hard/soft water; magnesium.)

water table — the upper limit of the portion of the ground wholly saturated with water. The water table separates two different layers of water in ground. The upper layer contains only a small amount of water for plant life; the bottom layer is the zone of saturation representing the water table. In between the two layers is a zone of aeration, where the spaces between the soil and rocks hold air. During periods of heavy rain the water moves into the zone of aeration, and the water table rises. In drought, the level of water in the saturation layer drops, indicating a corresponding drop in the water table.

wells — a water source that can be built by hand or power tools.

Bored wells, constructed by a hand-driven or power auger, are deeper and smaller in width as compared with dug wells. They are suited to sandy or soft-rock areas.

Drilled wells can be constructed through the hardest rock, to depths of up to 1,000 feet. Drilled wells are constructed with cable tool or rotary equipment.

Driven wells are the cheapest and quickest way to get water in a sandy area. Well construction is accomplished with a pointed object being forced into the ground. These wells go no deeper than 50 feet and are between 1¼ to 2 inches in diameter.

Dug wells usually go no deeper than 50 feet and range from 3 to 20 feet in diameter. They are more suited to geological areas that are sandy or have soft rock such as fractured limestone. Their major drawback is that they frequently bottom out during dry periods.

Jetted or hydraulic well drilling is most successful in sandy soils, but is not good in hard rock formations. The well hole is made by the force of a high velocity stream of water. The depth of these wells can reach 1,000 feet with a diameter ranging from 4 to 12 inches.

Appendix B: Water Supply Disinfection Chart*

A number of serious diseases such as hepatitis, typhoid fever, and amoebic dysentery, as well as a number of relatively minor disorders such as intestinal upsets, common diarrhea, and skin infections, have been traced to microorganisms that can travel through water supply systems. While waterborne diseases are less prevalent in the United States than in many lesser developed countries, several thousand cases are reported each and every year.

The best protection against waterborne disease is a naturally filtered water supply. Since few microorganisms can travel through more than a few dozen yards of fine sediment, a deep, properly grouted well in the middle of a large formation of fine sands, silts, and clays would be ideal. Springs or wells in rock or gravel formations often yield high quality water, but the absence of natural filtering materials makes them susceptible to contamination from surface sources. Lakes and rivers always should be the last choice because they're totally unprotected, but if they must be used the water can often be drawn from wells bored near the banks rather than directly from the water body (a German innovation). When assessing the safety of your water supply, always remember that laboratory tests of water quality can't reveal what may find its way into the supply during a flood, an epidemic, or some other chance occurrence.

Prepared by Lee Jaslow, Environmental Consultant, Baltimore, Maryland

Table B-1: Disinfection Scoreboard

Technique	Effectiveness	Reliability	Residual	By-Products	Pretreatment	Process Safety
Fine Filters	depends on size of filter openings: the finest filters remove most protozoa and bacteria but few viruses; suspended solids that may harbor microorganisms are also removed	buildup on filter reduces flow rate; bacterial breakthrough possible with some types	none	some filters can add bacteria; fiber filters may release fibers	filtration with coarse filter to remove larger particles that would clog fine filter; addition of bactericide to control growth of bacteria on fiber itself	no hazards
Ultrafilters or Reverse Osmosis	the best can remove nearly all microorganisms plus many other impurities	leaks are possible; frequent backwashing required; mineral buildup can clog filters	none	none	filtration with fine filter to remove larger particles; water should be low in minerals and have pH less than 7	no hazards

Technique	Effectiveness	Reliability	Residual	By-Products	Pretreatment	Process Safety
Chlorine (chlorine gas, calcium and sodium hypochlorite powders, hypochlorous acid solution)	can kill almost all microorganisms if adequate dosage applied for sufficient time	careful monitoring of dosing system required to assure constant application rate	leftover chlorine compounds	chlorinated organics and many other potentially toxic chlorinated compounds are possible	filtration with this fine filter and activated carbon to remove suspended solids that may harbor microorganisms as well as organic impurities that could react with chlorine to form undesirable by-products; pH should be less than 7.5	chlorine gas is exceedingly dangerous; other forms are safer but are corrosive and must be handled with care
Chloramines	weak and slow-acting; generally not effective	careful monitoring of dosing system required	leftover chloramines	chloramines may be considered undesirable; otherwise there are few by-products of concern	filtration with fine filter to remove suspended solids	chloramines are generated on-site from chlorine and ammonia, both of which are hazardous

Source: Prepared by Lee Jaslow, Environmental Consultant, Baltimore, Maryland.

Table B-1—*continued*

Technique	Effectiveness	Reliability	Residual	By-Products	Pretreatment	Process Safety
Chlorine Dioxide	can kill almost all microorganisms if adequate dosage applied for sufficient time	careful monitoring of dosing system required	leftover chlorine dioxide (may not last long)	chlorinated organics and other potentially toxic chlorinated compounds are possible but less abundant than with chlorine disinfection; chlorine dioxide and the chloride ions formed when it breaks down may be harmful	filtration with fine filter and activated carbon to remove suspended solids and organic impurities	chlorine dioxide is generated on-site from chlorine which is hazardous
Bromine (liquid bromine, hypobromous acid, bromine resin)	can kill almost all microorganisms if adequate dosage applied for sufficient time	liquid forms require careful monitoring of dosing system; bromine resin requires little attention	leftover bromine compounds	brominated organics and other potentially toxic brominated compounds are possible	filtration with fine filter and activated carbon to remove suspended solids and organic impurities	liquid bromine is exceedingly dangerous; often liquid forms are corrosive and must be handled carefully; bromine resin poses no hazards if kept dry

Technique	Effectiveness	Reliability	Residual	By-Products	Pretreatment	Process Safety
Bromamines	can kill almost all microorganisms if adequate dosage applied for sufficient time	careful monitoring of dosing system required	leftover bromamines	bromamines may be considered undesirable; otherwise there are few by-products of concern	filtration with fine filter to remove suspended solids	bromamines are generated on-site from bromine and ammonia, both of which can be hazardous
Iodine (iodine crystals, triiodide resin)	can kill almost all microorganisms if adequate dosage applied for sufficient time	little attention needed	leftover iodine and iodine compounds	water may become cloudy if iron bacteria present; iodinated organics and other potentially toxic iodine compounds may be formed but in much less abundance than with chlorine and bromine; iodine may be hazardous to some persons	filtration with fine filter to remove suspended solids; pH should be less than 7.5	iodine crystals must be handled with care; triiodide resin poses no hazards

Table B-1—*continued*

Technique	Effectiveness	Reliability	Residual	By-Products	Pretreatment	Process Safety
Ozone	can kill almost all microorganisms if adequate dosage applied for sufficient time	electrical parts of ozone-generating device require periodic inspection; automatic valves should be provided to shut off water flow in the event of malfunction	none under normal conditions but ozone may persist for some time at very low water temperatures	ozone will break down organics into simpler compounds, some of which are potentially toxic; use of ozone may increase bacterial growth in water lines	filtration with fine filter to remove suspended solids	ozone is a toxic and potentially explosive gas and should be handled with care; high-voltage electricity or ultraviolet radiation used to generate ozone may also pose hazards
Silver	weak and slow-acting; generally not effective	little attention needed	leftover silver	silver may be a health hazard in high concentrations	filtration with fine filters and activated carbon to remove suspended solids and organic impurities; water softening to remove minerals	no hazards

Technique	Effectiveness	Reliability	Residual	By-Products	Pretreatment	Process Safety
Heat	several minutes exposure to temperatures greater than 160° F will kill almost all disease-causing bacteria and most viruses but boiling for 30 min is needed to kill almost all microorganisms	heating equipment requires periodic inspection; automatic valves should be provided to shut off water flow in the event of malfunction	heat—until water cools	heat may be undesirable	none required	burn hazard
Ultraviolet Radiation	can kill almost all microorganisms if adequate dosage applied for sufficient time	lamp covers must be cleaned periodically; lamps must be replaced annually; automatic valves should be provided to shut off water flow in the event of malfunction	none	none	filtration with fine filter and activated carbon to remove suspended solids and organic impurities; iron removal if level is high	direct exposure to UV light is hazardous (well-designed systems prevent accidental exposure)

The second-best protection against waterborne disease is adequate disinfection. If your water supply could be easily contaminated, disinfection is essential; but even if it's safe you may want the security that disinfection offers. In order to discover which disinfectant is best for your supply, refer to Table B-1. Disinfection treatments are individually rated for:

effectiveness —measured by the ability of a disinfectant to remove nearly all potentially harmful microorganisms found in a given water supply. No disinfectant kills all microorganisms all of the time—sooner or later one or two always get by—and no disinfectant works exactly the same on waters of different chemical qualities. A disinfectant that acts only on certain species of microorganisms cannot be considered effective unless it's used in conjunction with a second disinfectant that takes care of the ones it misses.

reliability —ability of a disinfection process to operate for months or even years with little attention. No matter how effective a disinfectant may be, it's useless when a malfunction prevents proper application. Reliability is especially important in home-scale systems because homeowners tend to neglect routine maintenance.

residual —term used to describe a disinfection process that provides for continuing disinfection all the way to the point of use (spigot, showerhead, etc.). Residuals assure that any organisms surviving the main disinfection process won't be able to multiply in the pipes: without residuals, water reaching the tap may contain more microorganisms than were present prior to disinfection! Residuals also afford partial protection when external contaminants enter the water supply system during repair work or sudden pressure drops. The remarkable series of events that led to a hepatitis epidemic on the 1969 Holy Cross football team (see box) clearly illustrates the value of this type of protection.

by-products —substances found in disinfected water that weren't present prior to disinfection. They've become notorious ever since it was disclosed that chlorine used for disinfection can react with traces of organic matter found in water supplies to form chloroform and other substances believed to cause cancer. Although many by-products are almost certainly harmless, it makes sense to choose the disinfectant with the fewest by-products when all other factors are equal. Since, by their very nature, residuals are a type of by-product, a choice must often be made between the hazards posed by the residual and the hazards posed by not having the residual.

pretreatment —the removal of selected water impurities or the addition of selected chemicals prior to disinfection. Pretreatment is often used to adjust water chemistry so that the disinfectant will be more effective. It may also be employed to remove substances that would otherwise react with the disinfectant to form by-products. Excessive pretreatment needs complicate the disinfection process and make it very expensive.

process safety —a function of the hazards involved in manufacturing, transporting, storing, and using the required hardware and chemicals. Safety analyses should always be based on the principle that any accident that can happen, will happen.

Holy Cross Football Team Hepatitis Outbreak

During the last week of September and the first week of October 1969, an outbreak of infectious hepatitis, limited to members of the varsity football team, occurred. This illness resulted in the unprecedented cancellation of the remaining football schedule with only two games having been completed. . . . The attack rate was 93 percent.

The confinement of the illness to one group and the clustering of cases over a fifteen-day period suggested a common source, such as water or food, as the cause of the outbreak. . . . A review of the menus, plus interviews with the commissary personnel, failed to reveal any suspicious leads. The Jesuit faculty, who used the same dining facilities, were free from infection. . . .

Epidemiological investigation was then directed to the practice field. . . . The field is located at the summit of the campus overlooking the city. . . . The design of the water supply into the field provided water for both irrigation and drinking purposes. . . . The line has five interruptions along its course in the form of subsurface faucets used for irrigating the field. . . .

During the period of investigation, there occurred, by chance, a break in the water line 2 miles proximal to the college, and the forceful vacuum created was appreciated when the subsurface irrigation faucets were turned on. . . . If the subsurface irrigation faucets became submerged and the faucets were partially opened, subsurface water would siphon back into the system. . . .

The next two links in the chain of events responsible for the outbreak were the weather conditions during the first week of practice and the occurrence of five cases of infectious hepatitis involving one adult and four children who lived in a condemned dwelling immediately adjacent to the field. . . . The weather had been unusually warm with absolutely no rainfall. The children used the field as a playground and especially enjoyed bathing in water accumulated around the subsurface faucets. Frequently, the children would neglect to turn the water off and it would run through the night, submerging the faucets and creating large puddles.

The final link in the chain was a two-alarm fire which occurred in the early morning of August 29 in a tenement approximately 2 miles proximal to the college. The demand for water was sufficient to create a negative pressure at the practice field and to permit back siphonage of contaminated surface water into the system . . . .

The surface water is presumed to have become contaminated with infectious hepatitis virus excreted from the infected children, and this contaminated water was consumed by the football team on the morning of August 29. . . .

Source: Leonard J. Morse et al., "Holy Cross Football Team Hepatitis Outbreak," *Antimicrobial Agents and Chemotherapy* (Washington, DC: American Society for Microbiology, 1971), pp. 30–32.

Appendix C:
How to Visit Your Local
Water Treatment Plant*

Before drinking water flows from your faucet, it probably has been screened, pumped, stirred, settled, filtered, refiltered, stored in huge reservoirs, and finally pumped again through hundreds of miles of pipes, eventually finding its way to your home. The basis of all this activity is the water works plant. A trip through your local treatment plant gives you the opportunity to see and smell the water we all take for granted, and learn how it is transformed in the many purification processes.

Most treatment works are open to the public for visits or tours at any time as long as you call ahead. These plants are not the most popular of the public utilities for school tours and consequently it is worthwhile to arrange for your tour before you arrive. In large municipal systems, there often is one person designated to conduct the tours who may invite you to come at your convenience. However, in smaller communities, there may not even be one full-time operator but instead one town employee who checks on the water treatment operations periodically. Communities with populations of over 2,500 usually have one full-time employee responsible for the water works plant and your visit will have to fit into this person's schedule. Regardless of size,

---

*Prepared by Patricia M. Nesbitt, Environmental Consultant, Strasburg, Virginia

operators generally welcome their customers to come and view their work, and you can expect to have a pleasant and informative visit.

The water works plant is likely to be near the river, lake, or reservoir that is the town's raw water source. If groundwater is used, the site of the wells will locate the plant and pumping station. However, more often than not, groundwater is not treated at all (some communities use chlorine to provide bacteriological safety) and only the huge pumps will be visible.

Raw water first enters the huge pipes containing several screens that turn back fish, sticks, and other floating debris. Pumps lift the water to the plant from here for treatment. At the older plants, built when land was readily available near the water works, large reservoirs often surround the plant to provide natural settling and free pretreatment. The Washington, DC, Dalecarlia plant uses twenty-four-hour natural retention basins to obtain good removal rates of organic and settable solids. Most water works plants do not have these retention basins, but instead lift the water directly into the plant's coagulation, flocculation, and sedimentation basins. Here chemicals are added to the water to help the turbid matter, the organic constituents, and the bacterial contaminants coagulate into larger and heavier particles that will settle out of the water. Alum, iron salts, or various polymers are often used as coagulants. These chemicals attach themselves to the contaminants in the water, making them sticky so they adhere to each other. When they form floc—a white billowy material—the impurities settle on the bottom easily. These first steps in treatment will remove the majority of the turbidity, organic contaminants, and bacteria.

As the water moves out of the large sedimentation basins, it is usually chlorinated to kill off the remaining bacteria and to oxidize bad-taste and odor-causing organic compounds. The smell of the chlorine masks any remaining odors and almost instinctively you feel the water is safe enough to drink at this point—even before it is filtered.

Filters are next used to strain out the remaining turbid materials. They may consist of anthracite coal, 18 to 30 inches deep, often overlaid with several inches of sand. Older filters use only sand. Here again the appearances can be deceiving. These filters do not resemble any you know. Instead, you are surrounded with huge, sunken concrete chambers interconnected by a dozen 16-inch pipes all buried below concrete walkways. The room itself feels empty in contrast to the important activity going on below it, and your voice echoes in the hollowness. You look over the railings into the filters, themselves buried below 4 feet of water waiting to flow through the porous strainer. Examining intently, you can gradually make out the filter granules below and then begin to marvel that you are seeing through 4 feet of water. Remembering how the river initially looked, you will be struck by the amount of purification achieved in only these few steps.

Throughout the treatment process, the water's chemical quality is carefully monitored. Automatic readings are fed into the main control room, giving important information about chlorine residual, pH, alkalinity, temperature, oxygen content, and turbidity. Chemical adjustments may be needed to keep these parameters within the optimal range throughout the whole plant. Adjustment only at the end of the process as the water leaves the plant does not provide a margin of safety needed if that last chemical feeder should become defective.

These steps described so far—prechlorination, coagulation, flocculation, and sedimentation—are the most common treatments used for surface waters. After these, the water is chlorinated again in order to provide a chlorine residual throughout the distribution system, and then the water is discharged directly either into the pipes connected to your faucets or into the finished water storage reservoirs which are generally outside nearby the plant and/or at strategic locations in the city. These finished water reservoirs are almost always covered to protect the water from any contaminants. Sometimes pumps are located near the reservoirs to distribute water to its destinations, but most are gravity fed.

There are many other treatment methods used at central water works plants, all designed to control or remove undesirable qualities of water. Lime is used extensively to raise the water's pH and for corrosion control. Polyphosphates are also used for corrosion control. Aeration may be specifically designed into the treatment scheme if there are high iron, hydrogen sulfide, or carbon dioxide levels. In many areas the water is also softened using lime–soda ash treatments. Activated carbon filters may be used for taste and odor control or for organic chemical removal. Fluoride compounds often are added at the end of the treatment process.

From start to finish, the whole process may only take three hours. Longer retention in settling basins is preferred in plants where the space is available, but smaller plants can use coagulants and pressure filtration to speed up the process without sacrificing the end quality.

The trip water makes from the plant's end, into the storage basin, and finally to your tap, may take only ten minutes if you live near the plant, or possible twenty-four hours if you are at the far end of the distribution system.

Water plants usually operate round the clock, often with very small staffs. The night crew is significantly smaller than the day's, when most of the bacterial testing, chemical mixing, and management planning takes place. Very loud—even piercing—alarms are utilized in the control room to wake possible nodding night watchpeople, just in case some process needs immediate adjustments.

The laboratories in water works plants vary in size and complexities according to the amount and quality of water treated, and, as you might expect, also according to the wealth of the community. Large cities have fully equipped bacteriological and chemical lab facilities, whose continuous monitoring services provide high levels of quality control. Some labs may be staffed with several technicians, utilizing the latest gas chromatography and mass spectrophotometers to keep continuous watch on organic contaminants. At the other end of the scale is the small rural system which takes one water sample a month and sends it to the state lab for coliform tests.

Water treatment is a science, to be sure. But it also is very much an art that requires patience and craftsmanship to practice. New equipment and monitoring techniques are rapidly modernizing it with more scientific precision; yet, the water itself still defies such regimentation.

Appendix D:
Getting Your Water Tested

The following is a listing of state health authorities who'll provide information on how to test your water supply. They may also serve as information and referral services, answering individual questions and supplying names of commercial laboratories equipped for all types of testing. We have made every effort to verify names, addresses, and telephone numbers. These, however, may change without notice.

Generally speaking, it is less time consuming to begin by contacting your local water authority, county health department, or regional Department of Environmental Resources, approaching the state authorities directly for information on more complicated testing or if local help is unavailable.

Test prices can vary greatly from no charge to several hundred dollars. Many tests performed by state bureaus, however, are either free or require only a nominal fee. So when testing for contaminants, your best bet is to begin at home base.

Alabama   Department of Public Health
                     State Office Building
                     501 Dexter Avenue
                     Montgomery, AL 36130
                     (205) 832-3170

Alaska	Department of Health and Social Services Division of Public Health Environmental Health Section Pouch H 06 F Juneau, AK 99811 (907) 465-3120
Arizona	Department of Health Services 1740 West Adams Street Phoenix, AZ 85007 (602) 255-1024
Arkansas	Arkansas Department of Health 4815 West Markham Street Little Rock, AR 72201 (501) 661-2111
California	Public and Environmental Health Department of Health Services 714 P Street Sacramento, CA 95814 (916) 445-1102
Colorado	Colorado Department of Health 4210 East 11th Avenue Denver, CO 80220 (303) 320-8333
Connecticut	Department of Health 79 Elm Street Hartford, CT 06115 (203) 566-2279
Delaware	Division of Public Health Department of Health and Social Services Jesse S. Cooper Memorial Building Dover, DE 19901 (302) 678-4731

District of Columbia	Department of Environmental Services Environmental Health Administration Bureau of Air and Water Quality 5010 Overlook Avenue, SW Washington, DC 20032 (202) 767-7370
Florida	Health Program Office Department of Health and Rehabilitative Services 1323 Winewood Boulevard Tallahassee, FL 32301 (904) 488-4070
Georgia	Division of Physical Health 47 Trinity Avenue, SW Atlanta, GA 30334 (404) 656-4734
Hawaii	Hawaii Department of Health Environmental Protection and Health Services Division 1250 Punchbowl Street P.O. Box 3378 Honolulu, HI 96801 (808) 548-4682
Idaho	Department of Health and Welfare Division of Environment Water Quality Bureau Drinking Water Program 450 West State Street Boise, ID 83720 (208) 334-4253
Illinois	Department of Public Health 535 West Jefferson Street Springfield, IL 62761 (217) 782-5830

Indiana	Board of Health A-405 Health Building 1330 West Michigan Street Indianapolis, IN 46206 (317) 633-8400
Iowa	Iowa State Department of Health Health Engineering Section Lucas State Office Building East 12th and Walnut Streets Des Moines, IA 50319 (515) 281-5605
Kansas	Kansas Department of Health and Environment Forbes Field Topeka, KS 66620 (913) 862-9360
Kentucky	Department for Human Resources Bureau for Health Services Laboratory Services 275 East Main Street Frankfort, KY 40621 (502) 564-4446
Louisiana	Department of Health and Human Resources Office of Health Services and Environmental Quality P.O. Box 60630 New Orleans, LA 70160 (504) 568-5100
Maine	Bureau of Health Department of Human Services State House Augusta, ME 04333 (207) 289-3201

Maryland	Division of Water Supplies Department of Health and Mental Hygiene 201 West Preston Street Baltimore, MD 21201 (301) 383-3110
Massachusetts	Public Supplies only: Department of Public Health 600 Washington Street Boston, MA 02111 (617) 727-2700
Michigan	Department of Public Health 3500 North Logan Street P.O. Box 30035 Lansing, MI 48909 (517) 373-1376

or

Michigan Department of Public Health
Northern Peninsula Division
305 Ludington Street
Escanaba, MI 49829
(906) 786-6410

Minnesota	Department of Health 717 Delaware Street, SE Minneapolis, MN 55440 (612) 296-5227
Mississippi	Mississippi State Board of Health Division of Water Supply Bureau of Environmental Health P.O. Box 1700 Jackson, MS 39205 (601) 354-6646

Missouri	Division of Health Department of Social Services P.O. Box 570 Jefferson City, MO 65101 (314) 751-4330
Montana	Water Quality Bureau Department of Health and Environmental Sciences 200 Cogswell Building Lockey Street Helena, MT 59601 (406) 449-2544
Nebraska	Department of Health State Health Laboratory P.O. Box 2755 Lincoln, NB 68502 (402) 471-2122
Nevada	Division of Health Consumer Health Protection Services Kinkead Building 505 East King Street Carson City, NV 89710 (702) 885-4750
New Hampshire	New Hampshire Water Supply and Pollution Control Commission Division of Water Supply P.O. Box 95 Concord, NH 03301 (603) 271-3503

New Jersey

Department of Environmental Protection
Division of Water Resources
P.O. Box CN029
Trenton, NJ 08625
(609) 292-1637

New Mexico

Health and Environment Department
Environmental Improvement Division
Water Supply Unit
P.O. Box 968
Santa Fe, NM 87503
(505) 827-3201

New York

Department of Health
Tower Building, Room 482
Empire State Plaza
Albany, NY 12237
(518) 474-5577

North Carolina

Water Supply Branch
Environmental Health Section
Division of Health Services
Department of Human Resources
P.O. Box 2091
Raleigh, NC 27602
(919) 733-2321

North Dakota

North Dakota State Department of Health
State Capitol
Bismarck, ND 58505
(701) 224-2371

Ohio

Ohio Department of Health
General Environmental Health Services
246 North High Street
P.O. Box 118
Columbus, OH 43216
(614) 466-1390

Oklahoma	Water Quality Service
	Department of Health
	Northeast 10th and Stonewall Streets
	P.O. Box 53551
	Oklahoma City, OK 73152
	(405) 271-5205

Oregon
: Oregon State Health Division
  Public Health Laboratory
  1717 Southwest 10th Avenue
  Portland, OR 97201
  (503) 229-5882

Pennsylvania
: Bureau of Community Environmental Control
  Division of Public Water Supply
  P.O. Box 2063
  Harrisburg, PA 17120
  (717) 783-3795

Rhode Island
: Department of Health
  Cannon Health Building
  75 Davis Street
  Providence, RI 02908
  (401) 277-6867

South Carolina
: Department of Health and
  	Environmental Control
  Water Supply Division
  2600 Bull Street
  Columbia, SC 29201
  (803) 758-5544

South Dakota
: Department of Water and Natural Resources
  Joe Foss Building
  Pierre, SD 57501
  (605) 773-3754

Tennessee	Tennessee Department of Public Health 344 Cordell Hull Building Nashville, TN 37219 (615) 741-3111
Texas	Texas Department of Health Division of Water Hygiene 1100 West 49th Street Austin, TX 78756 (512) 458-7497
Utah	Utah State Division of Health 150 West North Temple Salt Lake City, UT 84113 (801) 533-6111
Vermont	Vermont Department of Health Division of Environmental Health 60 Main Street Burlington, VT 05402 (802) 862-5701
Virginia	Public Supply: Bureau of Water Supply Engineering Department of Health 109 Governor Street Richmond, VA 23219 (804) 786-1766 Private Supply: Community Health Services 109 Governor Street Richmond, VA 23219 (804) 786-3575

Washington

Health Services Division
Department of Social and Health Services
Mail Stop LD-11
Olympia, WA 98504
(206) 753-3466

West Virginia

Drinking Water Division
Department of Health
1800 Washington Street, East
Charleston, WV 25305
(304) 348-2971

Wisconsin

Division of Natural Resources
P.O. Box 7921
Madison, WI 53707
(608) 266-2621

Wyoming

Division of Health and Medical Services
Hathaway Building
c/o Public Health Laboratory Services
Cheyenne, WY 82002
(307) 777-7431

Index

Provincetown (MA) water supply,
contamination of, 8
Pumps, for wells, 184–85

Q

Quality
of groundwater, 54–55
maintaining, 205–11
physical signs of, 80–82
Quartz-mercury vapor lamp, 162

R

Radioactivity, 11–14
background, 11
testing for, 114–15
Rain, acid, 48
cisterns and, 188–89
Reverse osmosis, as water
treatment, 153–57, 232
Rice husks, as filter medium, 207
Rivers. *See* Surface water
Roof washer, for cistern, 190–91

S

Safe Drinking Water Act, 17–19
Salt, added by water softeners,
72–73
Salt, highway, as groundwater
contaminant, 58
Salt water
desalinization of, 205–6
as groundwater contaminant,
56
Sand filter, 135–36, 137–38
San Pellegrino Pure Mineral Table
Water, 125
Saratoga bottled water, 125
Sears Taste and Odor Filter, 142
Sea water. *See* Oceans; Salt water

Sediment, in water, 80–81. *See also*
Turbidity
Selenium, 102–3
in groundwater, 57
Septic tanks, contamination from,
55, 57
Sewage, discharged into surface
waters, 51
Silicon, heart disease and, 71–72
Silver, 6, 105
as disinfectant, 90, 141, 236
Skin cancer. *See* Cancer, skin
Smell. *See* Odor
Soda ash, to fight corrosion,
165–66
Sodium. *See also* Salt
hypertension and, 73
Soft water, 67–74
acidity of, 72
benefits of, 68
health effects of, 68–73
pipe corrosion and, 72, 93
sodium content of, 72–73
Soil and Health Foundation
hair testing by, 107
water testing by, 94–95
Soil type, effect of well, 178
Solids, suspended, as class of
pollutants, 6
Spa Reine bottled water, 123
Spinach, as nitrate source, 75
Springs, as primary water source,
194–98
Stills. *See* Distillation
Storage, of pure water, 149
Stress, lithium and, 101–2
Strokes, water hardness and, 69
Sugar metabolism, chromium and,
100
Sulfur, in groundwater, 55
Sulfur dioxide, 48
Surface water, 49–52